QUANTITATIVE
TECHNIQUES
FOR DECISION MAKING
IN CONSTRUCTION

QUANTITATIVE TECHNIQUES FOR DECISION MAKING IN CONSTRUCTION

S.L. Tang
Irtishad U. Ahmad
Syed M. Ahmed
Ming Lu

香港大學出版社
HONG KONG UNIVERSITY PRESS

Hong Kong University Press
14/F Hing Wai Centre
7 Tin Wan Praya Road
Aberdeen
Hong Kong

© Hong Kong University Press 2004

ISBN 962 209 705 7

British Library Cataloguing-in-Publication Data
A catalogue record for this book is available from the British Library.

Secure On-line Ordering
http://www.hkupress.org

Printed and bound by Condor Production Co. Ltd., Hong Kong, China

CONTENTS

PREFACE

Construction professionals all over the world are facing tremendous, ever-increasing challenges, both technical and managerial. They have to keep abreast with the latest techniques in both technology and management. Management techniques can be qualitative or quantitative. In the experience of the authors of this book, quantitative management techniques, unlike qualitative management, are seldom known to and rarely used by people in the construction industry for their decision making processes. Also, there are very few books written on the topic of quantitative management techniques for construction professionals. For these reasons, the authors believe this volume will be of value to people working in the field of construction.

This volume is intended to be a text book as well as a reference book on quantitative techniques for decision making in construction. The book is written to serve university undergraduates of construction-related programmes and postgraduate students undertaking construction management bridging courses. It teaches readers how to apply a number of quantitative techniques to assist their decision making processes when they are facing day-to-day problems at construction sites or in their offices.

The quantitative techniques covered in this book include analytic hierarchy process (AHP), decision theories, conditional probabilities and the value of information, inventory modeling, dynamic programming, Monte-Carlo simulation, CYCLONE simulation modeling, information systems and process of decision making in construction. Plenty of real life examples are used to illustrate the theories, arguments and calculations. It is written in simple and easy to understand language. Readers will benefit from the book even by self studying because all the topics are described and developed in a logical and organized manner, from basic to complex. Moreover, the most up-to-date information on the development of AHP, simulation modelling, information systems and process of decision making are covered.

The quantitative techniques discussed in this book do not include linear programming, integer programming and goal programming. These topics are contained in another book, *Linear Optimization in Applications*, written by S.L. Tang and published by the same publisher.

There are exercise questions at the end of each chapter. At the end of the book, solutions to selected questions (over half of the total number of exercise questions in all the chapters) are provided. This is particularly useful to those readers who intend to use this book for self studying. It is the hope of the authors that readers would find this book interesting and motivating, and would benefit from learning the quantitative techniques without a lot of guidance by experienced teachers.

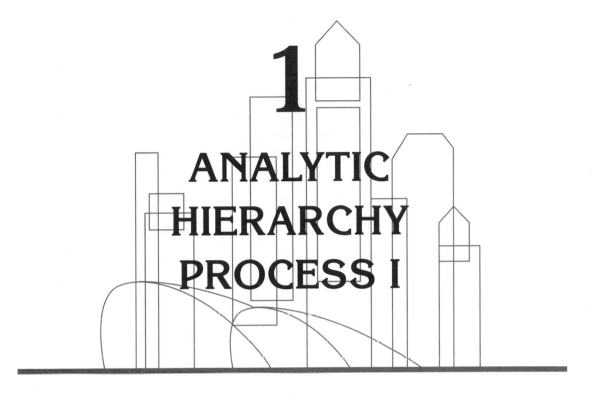

ANALYTIC HIERARCHY PROCESS I

1.1   What Is Analytic Hierarchy Process (AHP) ?

The "Analytic Hierarchy Process" (or AHP in short), a mathematical tool for management decision making, was introduced by Thomas L. Saaty (1977 and 1980). The mathematical technique is capable of handling a large number of decision factors and provides a systematic procedure of ranking many decision variables. It is a decision analysis technique which can be very useful in construction management. This chapter will firstly give a brief description of the theory of AHP. Cases will then be used to illustrate how this analysis technique can be applied in the field of construction.

1.2   Mathematical Theory of AHP

1.2.1 We will use the selection of tenders as an example in explaining the AHP theory. Suppose that there are n factors in considering whether a tender should be accepted or not. These n factors have different importance contributing to the acceptance or unacceptance of the tender. In assessing the importance of each factor, pairwise comparisons are used so that any one factor is not compared to all other factors simultaneously but rather *one at a time*. For an easy explanation let us take n = 3 in the following example.

Three factors; 1, 2 and 3.

Factor 1 is twice as important as Factor 2.

Factor 2 is three times as important as Factor 3.

Factor 1 is six times as important as Factor 3.

Then, these scores can be entered into a matrix A as follows:

$$A = \begin{array}{c|ccc} & 1 & 2 & 3 \\ \hline 1 & 1 & 2 & 6 \\ 2 & 1/2 & 1 & 3 \\ 3 & 1/6 & 1/3 & 1 \end{array}$$

Matrix A is called the "reciprocal matrix" because all its lower triangular elements are equal to the reciprocal of its upper triangular elements. The maximum eigenvalue λ of matrix A can be found by solving

$$| A - \lambda I | = 0$$

The eigenvector X, corresponding to the maximum eigenvalue λ, which satisfies that $AX = \lambda X$, is found to be:

$$X = \begin{bmatrix} 0.6 \\ 0.3 \\ 0.1 \end{bmatrix} \begin{array}{l} - \text{Factor 1} \\ - \text{Factor 2} \\ - \text{Factor 3} \end{array}$$

The higher the value of the element in X, the more important the factor is.

1.2.2 Readers may find it difficult to understand why the resultant eigenvector can represent importance. The following gives a brief explanation. Details of it can be found in the work of Saaty (1977 and 1980).

Suppose a set of n objects are to be compared in pairs and their individual importance (assumed known) are denoted by $x_1, x_2, \ldots, x_n$, then the pairwise comparison may be represented by a (n x n) square reciprocal matrix A as follows:

$$A = \begin{array}{c|cccccc} & 1 & 2 & \ldots & \ldots & \ldots & n \\ \hline 1 & 1 & x_1/x_2 & \ldots & \ldots & \ldots & x_1/x_n \\ 2 & x_2/x_1 & 1 & & & & x_2/x_n \\ \vdots & \vdots & & & & & \vdots \\ \vdots & \vdots & & & & & \vdots \\ \vdots & \vdots & & & & & \vdots \\ n & x_n/x_1 & x_n/x_2 & \ldots & \ldots & \ldots & 1 \end{array}$$

If a column vector X is defined such that x_1, x_2, ..., x_n (i.e. the individual importance of objects 1, 2, ..., n) are the elements of X such that

$$X = \begin{bmatrix} x_1 \\ x_2 \\ \vdots \\ \vdots \\ x_n \end{bmatrix}$$

then,

$$AX = \begin{bmatrix} 1 & x_1/x_2 & \cdots & x_1/x_n \\ x_2/x_1 & 1 & & x_2/x_n \\ \vdots & & & \vdots \\ \vdots & & & \vdots \\ x_n/x_1 & x_n/x_2 & \cdots & 1 \end{bmatrix} \begin{bmatrix} x_1 \\ x_2 \\ \vdots \\ \vdots \\ x_n \end{bmatrix}$$

$$= \begin{bmatrix} x_1 + x_1 + \ldots\ldots + x_1 \\ x_2 + x_2 + \ldots\ldots + x_2 \\ \vdots \\ \vdots \\ x_n + x_n + \ldots\ldots + x_n \end{bmatrix}$$

$$= n \begin{bmatrix} x_1 \\ x_2 \\ \vdots \\ \vdots \\ x_n \end{bmatrix}$$

$$= nX$$

Therefore, AX = nX, and n is, by definition, an eigenvalue of matrix A. The matrix X is the solution eigenvector of $AX = \lambda X$ for taking $\lambda = n$. n is, by Perron-Frobenius Theorem, the maximum eigenvalue of A. This explains why the eigenvector contains the ranking of the importance of the n objects.

1.2.3 The above theory is based on the assumption that all entries of the matrix A are consistent which means that $(a_{ij})(a_{jk}) = a_{ik}$. The example in Section 1.2.1 is a consistent case. Such perfect consistency is possible only if we can construct matrix A based on the weightings of individual objects (i.e. x_1, x_2, ..., x_n). However, in the application of AHP, this will not be the case, that is, one can only construct matrix A first by pairwise comparisons and then find out the values of x_1, x_2, ..., x_n. This will create inconsistency in the reciprocal matrix A. Just take a new example of three football teams: if team 1 beats team 2 by 2:1 and team 2 beats team 3 by 3:1, it is not necessarily the case that team 1 will beat team 3 by 6:1 (this is exactly the case in the example in Section 1.2.1), and if not, inconsistency occurs.

When inconsistency occurs, the problem AX = nX becomes $AX = \lambda_{max} X$. For the reciprocal matrix A, λ_{max} will not be equal to n. It has been proved by Saaty that λ_{max} is closer to n when matrix A is closer to consistency. A is consistent if and only if λ_{max} = n. λ_{max} is always greater than n if A is inconsistent. The further λ_{max} is from n, the more the inconsistent the matrix is.

Let us now modify the example in Section 1.2.1 to an inconsistent case as follows:

Three factors: 1, 2 and 3.
Factor 1 is twice as important as Factor 2.
Factor 2 is three times as important as Factor 3.
Factor 1 is four times as important as Factor 3.

The reciprocal matrix is therefore written as:

$$A = \begin{bmatrix} 1 & 2 & 4 \\ 1/2 & 1 & 3 \\ 1/4 & 1/3 & 1 \end{bmatrix}$$

The maximum eigenvalue is 3.018 and the corresponding eigenvector X (or called **priority vector**) is:

$$x = \begin{bmatrix} 0.56 \\ 0.32 \\ 0.12 \end{bmatrix}$$

Readers can now see the difference of this priority vector and the priority vector in Section 1.2.1.

It is important to note that the summation of all the elements in a priority vector is equal to 1 (see the two previous examples). This is called "normalization" of the priority vector. The normalization is necessary in order to ensure uniqueness of the vector. The two priority vectors we have seen are said to be **normalized additively**, that is, the elements add up to 1. There are, however, cases that need a priority vector to be **normalized multiplicatively**. We will see such examples in Section 2.3 in Chapter 2. The evaluation of the maximum eigenvalue of a n × n matrix and hence its corresponding eigenvector (i.e. the priority vector) can be easily found in many mathematics books and software on the market.

Table 1.1 shows the scales of 1 to 9 as recommended by Saaty for inputting values into the reciprocal matrix.

Intensity of relative importance	Definitions	Explanation
1	Equal importance	Two activities contribute equally to the objectives
3	Moderate importance of one over another	Experience and judgement slightly favoured one activity over another
5	Essential or strong Importance	Experience and judgement strongly favoured one activity over another
7	Demonstrated importance	An activity is strongly favoured and its dominance is demonstrated in practice
9	Extreme importance	The evidence favouring one activity over another is of the highest possible order of affirmation
2,4,6,8	Intermediate values between the two adjacent judgements	When compromise is needed

Table 1.1 Scales of 1 to 9 for pairwise comparisons (Saaty 1977)

1.3 Three Levels of Hierarchy

So far, only how n objects are ranked based on a single objective has been discussed. Now, a hierarchy of 3 levels (Fig. 1.1) is looked into. The first level has a single goal (or principal objective). The second level has m subordinate objectives (or factors) and their rankings are derived from pairwise comparisons based on the goal of the first level. The third level has n objects (or tenderers) which are to be ranked. The problem is to determine how well the objects meet the goal through the intermediate second level of subordinate objectives. The procedures are described below.

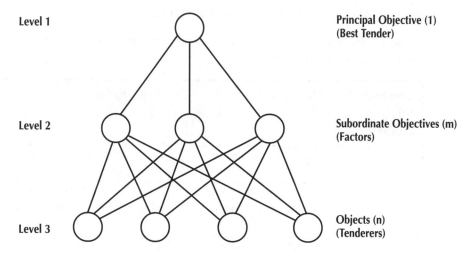

Fig. 1.1 A 3-level hierarchy structure

Step 1   Construct a reciprocal matrix B of dimension m x m with pairwise comparisons of the factors with respect to the principal objective as the elements of B. Then find the priority vector of matrix B and denote it by X_b.

Step 2   Since there are n tenderers to be ranked, a total of m reciprocal matrices A_i (i = 1, 2, ..., m) of size n × n are to be formed, each of which consists of elements of pairwise comparisons of the tenderers with respect to a single factor as objective, that is:

$$A_1 = \begin{array}{c|cccc} & 1 & 2 & ... & n \\ \hline 1 & & & & \\ 2 & & & & \\ : & & & & \\ : & & & & \\ n & & & & \end{array}$$ Using factor 1 as objective

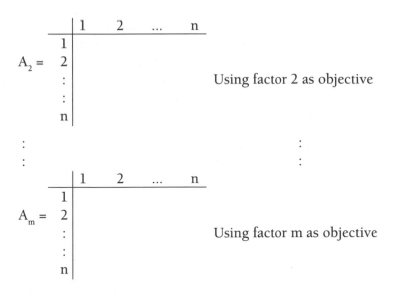

Step 3   If X_i (i = 1, 2, ..., m) is the priority vector of the corresponding A_i, then an n x m composite matrix C can be formed by taking X_i as columns in C in sequence such that:

$$C = [X_1, X_2, ..., X_m]$$

Step 4   The resultant priority vector, X_c, is the result of the multiplication of matrices C and X_b, that is:

$$X_c = C \times X_b$$

From the result of X_c, the ranking of the tenderers can be obtained. These 4 steps will be further illustrated in the next section.

1.4   **Evaluation of Tenders**

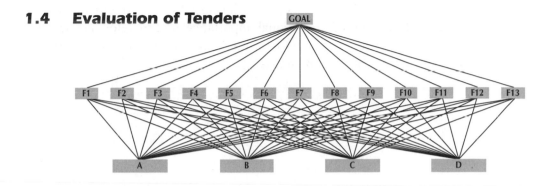

i.e., m=13 & n=4

Fig. 1.2 3-level hierarchy for evaluation of tenders

A tender evaluation example is shown as a 3-level hierarchy AHP problem (Fig. 1.2). Four tenders, A, B, C and D, are to be evaluated. They are from four different consulting firms who wish to bid to undertake the planning, design and supervision of construction for a construction project. So, four tender proposals have been submitted to the client. There are thirteen factors (under five headings) to be considered by the client. These factors are:

Consultant's Experience
F1:   Relevant experience and knowledge.
Response to The Brief
F2:   Understanding of objectives.
F3:   Identification of key issues.
F4:   Appreciation of project constraints and special requirements.
F5:   Presentation of innovative ideas.
Approach to Cost-Effectiveness
F6:   Examples and discussion of past projects to demonstrate the consultant's will and ability to produce cost-effective solutions.
F7:   Approach to achieve cost-effectiveness on this project.
Methodology and Work Programme
F8:   Technical approach.
F9:   Work programme and project implementation programme.
F10: Arrangements for contract management and site supervision.
Staffing
F11: Organization structure.
F12: Relevant experience and qualification of key staff.
F13: Responsibilities and degree of involvement of key staff.

The matrix B (see step 1 of Section 1.3) is shown in Fig. 1.3. Its elements are pairwise comparisons of the factors with respect to the principal objective. For example, F1 and F2 are of equal importance, and therefore element 1-2 of matrix B is entered as 1. F3 is of moderate importance over F10, and therefore element 3-10 is entered as 3.

Goal	F1	F2	F3	F4	F5	F6	F7	F8	F9	F10	F11	F12	F13
F1	1	1	1	1	1	1	1/5	1/7	1/3	3	1	1/7	1/7
F2		1	1	1	1	1	1/5	1/7	1/3	3	1	1/7	1/7
F3			1	1	1	1	1/5	1/7	1/3	3	1	1/7	1/7
F4				1	1	1	1/5	1/7	1/3	3	1	1/7	1/7
F5					1	1	1/5	1/7	1/3	3	1	1/7	1/7
F6						1	1/5	1/7	1/3	3	1	1/7	1/7
F7							1	1/5	4	7	5	1/5	1/5
F8								1	6	9	7	1	1
F9	All lower triangular elements								1	5	3	1/6	1/6
F10	$=a_{ji}$									1	1/3	1/9	1/9
F11	$=1/a_{ij}$										1	1/7	1/7
F12												1	1
F13													1

Fig. 1.3 Matrix B for tender evaluation

The priority vector X_b of matrix B is:

$$X_b = \begin{bmatrix} 0.0258 \\ 0.0258 \\ 0.0258 \\ 0.0258 \\ 0.0258 \\ 0.0258 \\ 0.1074 \\ 0.2136 \\ 0.0586 \\ 0.0126 \\ 0.0258 \\ 0.2136 \\ 0.2136 \end{bmatrix}$$

The next step is to construct 13 matrices of size 4×4, each of which consists of elements of pairwise comparisons of the 4 tenders with respect to a single factor as objective. It is impossible to show all the 13 matrices here and only A_1, A_2 and A_{13} are shown below:

$$A_1 = \begin{array}{c|cccc} & A & B & C & D \\ \hline A & 1 & 1 & 1 & 2 \\ B & 1 & 1 & 1 & 2 \\ C & 1 & 1 & 1 & 2 \\ D & 1/2 & 1/2 & 1/2 & 1 \end{array}$$ Using Factor 1 as objective

$$A_2 = \begin{array}{c|cccc} & A & B & C & D \\ \hline A & 1 & 1/2 & 2 & 1 \\ B & 2 & 1 & 4 & 2 \\ C & 1/2 & 1/4 & 1 & 1/2 \\ D & 1 & 1/2 & 2 & 1 \end{array}$$ Using Factor 2 as objective

$$\vdots \qquad\qquad\qquad\qquad\qquad\qquad \vdots$$

$$A_{13} = \begin{array}{c|cccc} & A & B & C & D \\ \hline A & 1 & 2 & 5 & 5 \\ B & 1/2 & 1 & 3 & 3 \\ C & 1/5 & 1/3 & 1 & 1 \\ D & 1/5 & 1/3 & 1 & 1 \end{array}$$ Using Factor 13 as objective

The priority vectors, $X_1, X_2, \ldots, X_{13}$, of the matrices, $A_1, A_2, \ldots, A_{13}$ respectively are:

$$X_1 = \begin{bmatrix} 0.2857 \\ 0.2857 \\ 0.2857 \\ 0.1428 \end{bmatrix} \qquad X_2 = \begin{bmatrix} 0.2222 \\ 0.4444 \\ 0.1111 \\ 0.2222 \end{bmatrix} \qquad \ldots\ldots X_{13} = \begin{bmatrix} 0.5183 \\ 0.2839 \\ 0.0989 \\ 0.0989 \end{bmatrix}$$

Hence, a composite matrix C can be formed as follows. Note that the 1st, 2nd and 13th columns of C are X_1, X_2 and X_{13} respectively.

$$C = \begin{bmatrix} 0.2857 & 0.2222 & \ldots\ldots & 0.5183 \\ 0.2857 & 0.4444 & \ldots\ldots & 0.2839 \\ 0.2857 & 0.1111 & \ldots\ldots & 0.0989 \\ 0.1428 & 0.2222 & \ldots\ldots & 0.0989 \end{bmatrix}$$

The ranking of the four tenders = C x X_b

$$= \begin{bmatrix} 0.3548 \\ 0.2599 \\ 0.0851 \\ 0.3002 \end{bmatrix}$$

Therefore, the best tender is A, which scores 0.3548; the second best tender is D, which scores 0.3002; and so on.

1.5   Four Levels Hierarchy — Tender Evaluation

A 3-level hierarchy has just been discussed. The following will illustrate a more complicated tender evaluation example in which a 4-level hierarchy is involved (Tang, 1995).

There are four international contractors tendering for a large civil engineering design-and-build contract involving earth work, road work, tunnel work, building work and E&M work. Each of these works forms a part of the contract but each is significant enough to form a contract by itself if it is not a large turn-key international contract requiring high standard of workmanship and reliability. Hence, the selection of tender must be carried out with exceptional care. Fig. 1.4 shows a 4-level hierarchy, following which the client evaluates the tenders.

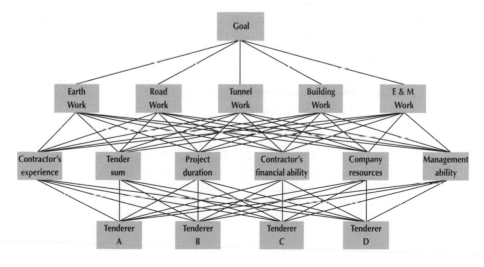

Fig. 1.4 4-level hierarchy for evaluation of tenders

The procedures of evaluating a 4-level hierarchy are described as follows:

Step 1   Construct a reciprocal matrix P of dimension 5×5 with pairwise comparisons of the different works with respect to the principal objective as the elements of P. The higher the percentage of the work in the contract, the more important that work is. Then find the priority vector of matrix P and denote it by P_b.

Step 2   Construct five reciprocal matrices Q_i (i = 1, 2, 3, 4, 5) of size 6×6, each of which consists of elements of pairwise comparisons of the six factors with respect to a particular work as objective.

Step 3   If X_i (i = 1, 2, 3, 4, 5) is the priority vector of the corresponding Q_i, then a 6×5 composite matrix R is formed by taking X_i as columns in R in sequence such that:

$$R = [X_1, X_2, X_3, X_4, X_5]$$

Step 4   Construct six reciprocal matrices S_i (i = 1, 2, 3, 4, 5, 6) of size 4×4, each of which consist of elements of pairwise comparisons of the four tenderers with respect to a single factor as objective.

Step 5   If Y_i (i = 1, 2, 3, 4, 5, 6) is the priority vector of the corresponding S_i, then a 4×6 composite matrix T can be formed by taking S_i as columns in T in sequence such that:

$$T = [Y_1, Y_2, Y_3, Y_4, Y_5, Y_6]$$

Step 6   The resultant priority vector, X_c, is the result of the multiplication of matrices T, R and P_b, that is:

$$X_c = T \times R \times P_b$$

Similar to the previous example, the ranking of the tenderers can be known after the result of X_c is obtained.

1.6   **More Examples on Applications of AHP**

Example 1.1   A 3- level hierarchy decision problem

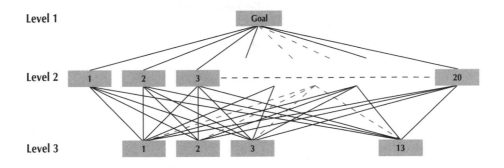

Level 1   Goal

Level 2   1   2   3   –   20

Level 3   1   2   3   13

Level 1

Goal: Select the best wastewater treatment alternative

Level 2

Factors for consideration:
1. Sewage flow
2. Influent/Effluent standard
3. Size of site
4. Nature of site
5. Land cost
6. Local money for construction
7. Foreign money component for construction
8. Local skill for construction
9. Community support
10. Power source
11. Availability of local material
12. Cost of operation and maintenance
13. Professional skill available for operation and maintenance
14. Local technical skill available for operation and maintenance
15. Administration set-up
16. Training
17. Professional ethics
18. Climate
19. Local water-borne diseases
20. Endemic vector-borne (water-related) diseases

Level 3

Wastewater treatment alternatives:
1. Stabilization ponds
2. Fully aerated lagoons + Secondary settlement
3. Fully/Partially aerated lagoons
4. Simple percolating filtration
5. Modified percolating filtration
6. Conventional activated sludge process
7. Deep-shaft/High-purity oxygen processes
8. Primary settlement
9. Land application
10. Rotating biological contactors
11. Oxidation ditches
12. Package activated sludge plants
13. Package high-purity oxygen plants

Example 1.2 A 4-level hierarchy decision problem

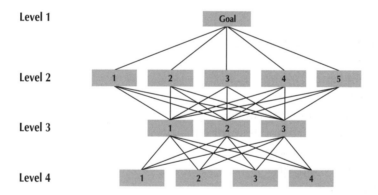

Level 1

Goal: Select the optimal alignment for a road project

Level 2

Design aspects:
1. Project sum
2. Project duration
3. Construction risks
4. Operation and maintenance
5. Environmental impact

Level 3

Construction methods:
1. Tunnel
2. Bridge
3. Immersed tube

Level 4

Alignment alternatives:
1. Alignment 1
2. Alignment 2
3. Alignment 3
4. Alignment 4

Example 1.3 Another 4-level hierarchy decision problem

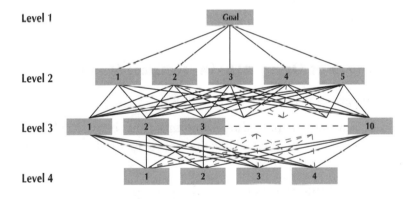

Level 1

Goal: Select the most appropriate contractual arrangement for a construction project

Level 2

Components of the project:

1. Earthwork
2. Roadwork
3. Drainage and sewerage work
4. Building work
5. Utility work

Level 3

Factors affecting the choice of contractual agreement:
1. Project definition
2. Owner preferences
3. Public laws
4. Current market conditions
5. Project location
6. Project financing
7. Schedule
8. Assumption of risks
9. Scope of work
10. Duration of work

Level 4

Contractual arrangement alternatives
1. Cost plus contract
2. Target price contract
3. Unit price contract
4. Lump sum contract

1.7   Advantages of Using AHP in Decision Making

The advantages of using AHP are as follows:

1.7.1 It provides a systematic procedure for comparisons between objects under a large number of factors. It facilitates the employment of subjective weighing of objects based on experience.

1.7.2 Besides quantifiable factors (e.g. tender sum, project duration, etc.), the method enables the consideration of unquantifiable/subjective factors which are important in decision making processes.

1.7.3 The size and the complexity of a problem can be broken down into small items (or called clusters) for analysis. For the number of levels of the hierarchy is flexible depending on the size and the requirements of the problem.

1.7.4 The resultant priority vector obtained from this method can give an indication of how much one object is better than another. This can hardly be achieved by intuition alone.

Exercise Questions

Question 1

Explain the consistency and inconsistency of a reciprocal matrix in the Analytic Hierarchy Process (AHP). How is the matrix's largest eigenvalue (λ_{max}) related to the consistency?

In the application of AHP, what type of reciprocal matrix (consistent or inconsistent) is usually used? Why?

Question 2

Construct a 5-level hierarchy problem related to construction, the 1st level being the goal and the 5th level being the alternatives for selection.

2

ANALYTIC HIERARCHY PROCESS II

2.1 The Controversies of AHP

2.1.1 Right and Left Eigenvector Approaches

In Chapter 1, we have discussed the approach that Saaty proposed, which is called **right eigenvector approach**. The right eigenvector approach means that the priority vector X is calculated based on the equation $AX = \lambda_{max} X$.

However, we can also calculate the priority vector using the **left eigenvector approach.** The left eigenvector approach means that the priority vector is calculated based on

$$(AX)^T = (\lambda_{max} X)^T$$

$$\text{ie. } X^T A^T = \lambda_{max} X^T$$

Johnson, Beine and Wang (1979), two years after Saaty proposed his AHP theory, discovered that the priority vectors calculated from a same reciprocal matrix using the right eigenvector approach and the left eigenvector approach may have disagreed results. Let us look at the following example.

Example 2.1

Find the priority vectors of matrix A using the right eigenvector approach and the left eigenvector approach.

$$A = \begin{bmatrix} 1 & 1 & 3 & 9 & 9 \\ 1 & 1 & 5 & 8 & 5 \\ 1/3 & 1/5 & 1 & 9 & 5 \\ 1/9 & 1/8 & 1/9 & 1 & 1 \\ 1/9 & 1/5 & 1/5 & 1 & 1 \end{bmatrix}$$

Solution

Using the right eigenvector approach, we solve $AX = \lambda_{max} X$. We obtain that

$$X = \begin{bmatrix} 0.366 \\ 0.389 \\ 0.167 \\ 0.035 \\ 0.042 \end{bmatrix} \begin{matrix} - 2^{nd} \\ - 1^{st} \\ - 3^{rd} \\ - 5^{th} \\ - 4^{th} \end{matrix}$$

Using the left eigenvector approach,

$$A^T = \begin{bmatrix} 1 & 1 & 1/3 & 1/9 & 1/9 \\ 1 & 1 & 1/5 & 1/8 & 1/5 \\ 3 & 5 & 1 & 1/9 & 1/5 \\ 9 & 8 & 9 & 1 & 1 \\ 9 & 5 & 5 & 1 & 1 \end{bmatrix}$$

We solve $X^T A^T = \lambda_{max} X^T$ and obtain that

$$X^T = \begin{bmatrix} 0.039 & 0.043 & 0.105 & 0.458 & 0.355 \end{bmatrix}$$
$$\quad\quad 1^{st} \quad\quad 2^{nd} \quad\quad 3^{rd} \quad\quad 5^{th} \quad\quad 4^{th}$$

In simple words, the right eigenvector approach is based on the pairwise comparison of elements of how one element is **better** than another, while the left eigenvector approach is based on the pairwise comparison of how one element is **worse** than another.

It can be seen that the ranks of the lst object and the 2nd object are reversed in the two approaches. Readers should note that, unlike the right eigenvector approach, the smaller the value in the eigenvector, the higher the ranking of the object is when we use the left eigenvector approach. The results show that the two approaches give contradictory rankings.

John, Beine and Wang (1979) argued that there is no reason to believe that utilization of a right eigenvector (as proposed by Saaty) yields a better result than the left. They experimented 364 randomly generated reciprocal matrices of size n=6 and there were 195 such ranking reversals between left and right eigenvectors.

Three years later, Vargas (1982) reported the finding of his research that if the **consistency index** of a reciprocal matrix is not larger than 10% then the result of the right eigenvector approach will be sufficiently reliable. As we have seen in Section 1.2.3 that the further λ_{max} is from n for an n x n reciprocal matrix, the more the inconsistent the matrix is. The consistency index is indicated by the difference of λ_{max} and n, and is defined as follows:

$$\text{Consistency index} = (\lambda_{max} - n) \div (n - 1)$$

Vargas' conclusion is that when a decision is made with the help of Analytic Hierarchy Process, the pairwise comparison reciprocal matrix must be tested first to see whether or not its consistency index is within 0 to 10% range, and if it is, then the priorty vector obtained using the right eigenvector approach is sufficiently reliable. Therefore, in the application of AHP as described in Chapter 1, we need to first test the consistency index of a reciprocal matrix and see if it is within 10%, and then find its priority vector (right eigenvector approach) for any decision making processes using AHP.

2.1.2 Modified AHP Proposed by Donegan et al.

Later, Donegon, Dodd and McMaster (1992, 1995) in their papers claimed that the right and left eigenvector inconsistency problem of Saaty's AHP (SAHP) can be effectively reduced by a new method which was known as Modified AHP (MAHP). The prime criticism of the SAHP by Donegon *et al.* was the scales used, i.e. 1 to 9, as recommended by Saaty. They pointed out that the ratios of the relative importance of pairs of elemental issues were on a ratio scale and hence multiplicative, in which the binary operations + and x in matrix manipulations implied that the rules of ordinary arithmetic apply. As stated in their paper, the SAHP is based on a set of positive integers and the corresponding multiplicative inverse (reciprocals) and hence a scale which is partly linear and partly harmonic. The method therefore appears to be "unmathematical". In order to solve the problem, they suggested a set of new scales by mappings of the Saaty's scales as follows.

If the matrix entries are mapped into R (R is a function of θ and ϕ) using

$$\theta : t \rightarrow \tanh^{-1}\left(\frac{t-1}{9}\right)$$

then the co-domain is additive. However, if the matrix is to be a positive reciprocal matrix, its entries must belong to a multiplicative co-domain and are determined by the mapping

$$\phi : t \rightarrow exp\,(\theta(t))$$

After the mappings, the original Saaty's scales of 1 to 9 will be mapped into another set of numerical values as shown in Table 2.1.

Original Saaty's scales	Calculations		Modified scales after mappings
	θ	ϕ	
1	0.000	1.000	1.000
2	0.112	1.118	1.118
3	0.226	1.254	1.254
4	0.347	1.414	1.414
5	0.478	1.612	1.612
6	0.626	1.871	1.871
7	0.805	2.236	2.236
8	1.040	2.828	2.828
9	1.417	4.123	4.123

Table 2.1   Mappings from Saaty's scales to modified scales

Example 2.2

Use MAHP to find out the right and left priority vectors of matrix A given in Example 2.1.

Solution

After mappings, matrix A becomes:

$$A = \begin{bmatrix} 1 & 1 & 1.254 & 4.123 & 4.123 \\ & 1 & 1.612 & 2.828 & 1.612 \\ & & 1 & 4.123 & 1.612 \\ \text{For lower triangular} & & 1 & 1 \\ \text{elements, } a_{ji} = \dfrac{1}{a_{ij}} & & 1 & 1 \end{bmatrix}$$

Solving $AX = \lambda_{max} X$ (right eigenvector approach) we obtain

$$X = \begin{bmatrix} 0.321 \\ 0.263 \\ 0.224 \\ 0.081 \\ 0.111 \end{bmatrix} \begin{matrix} - 1^{st} \\ - 2^{nd} \\ - 3^{rd} \\ - 5^{th} \\ - 4^{th} \end{matrix}$$

Using the left eigenvector approach, A^T becomes:

$$A^T = \begin{bmatrix} 1 & & \text{For upper triangular} \\ 1 & 1 & \text{elements, } a_{ji} = \dfrac{1}{a_{ij}} \\ 1.254 & 1.612 & 1 \\ 4.123 & 2.828 & 4.123 & 1 & 1 \\ 4.123 & 1.612 & 1.612 & 1 & 1 \end{bmatrix}$$

Solving $X^T A^T = \lambda_{max} X^T$, we obtain

$$X^T = \begin{bmatrix} 0.093 & 0.116 & 0.137 & 0.374 & 0.280 \end{bmatrix}$$
$$\begin{matrix} \mid & \mid & \mid & \mid & \mid \\ 1^{st} & 2^{nd} & 3^{rd} & 5^{th} & 4^{th} \end{matrix}$$

So, no rank reversal is observed between the two eigenvector approaches and it appears that the inconsistency problem in the SAHP can be reduced by the MAHP.

2.2   A Comparison of SAHP and MAHP

2.2.1 Research Done by Tung and Tang

Tung and Tang (1998), a few years after Donegon *et al.* published their papers about MAHP, attempted to compare SAHP and MAHP using 42 models comprising 294 reciprocal matrices. 21 models, comprising 147 reciprocal matrices, were classified as models for "pre-arranged order of objects". The remaining 21 models, comprising 147 reciprocal matrices too, were classified as models for "less obvious order of objects". Example 2.3 explains these two categories of models.

Example 2.3

Suppose a decision has to be made using AHP for the following 3-level hierarchy.

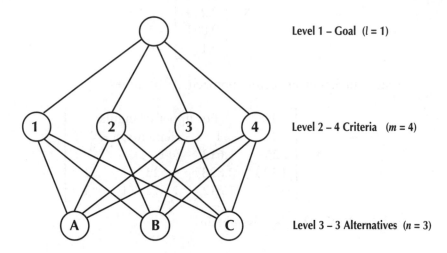

Level 1 – Goal ($l = 1$)

Level 2 – 4 Criteria ($m = 4$)

Level 3 – 3 Alternatives ($n = 3$)

In this hierarchy, the goal (l) is situated at Level 1 just like any other hierarchies. At Level 2, there are four ($m = 4$) criteria, 1, 2, 3 and 4. At the lowest level, Level 3, three ($n = 3$) alternatives, A, B and C are for prioritization. The objective is to select the best alternative through the 4 criteria so as to achieve the overall goal. Use both AHP methods (i.e. SAHP and MAHP), each to be tested with the right and left eigenvector approaches, to analyze problems of "pre-arranged order of objects" and "less obvious order of objects".

Solution

Pre-arranged order of objects

One way to study the right and left inconsistency problem is to adopt the method of "pre-arranged order of objects" in which the result of each level of the hierarchy as well as the overall ranking of all the alternatives were assumed known prior to carrying out any analysis work. Then, by applying the two AHP methods (i.e. SAHP and MAHP), each with both right and left approaches, the results can be inspected and checked against the predetermined order. If any one of the AHP methods fails to yield the prescribed results, the correctness of that method will be in doubt.

Step 1 - Rank Level 2 (Criteria) with respect to Level 1 (Goal)

Using pre-arranged order of objects, with respect to the goal at Level 1, the ranking order of the four criteria is, for example, Criterion1, then Criterion 2, then Criterion 3 and then Criterion 4.

Step 2 - Rank Level 3 (Alternatives) with respect to Level 2 (Criteria)

Likewise, in order to have a pre-arranged order of the three alternatives, the order of the alternatives are pre-set as, for example, Alternative A, then Alternative B and then Alternative C. Such an order is to be used for all the three criteria.

Less obvious order of objects

Unlike the method of pre-arranged order of objects, the order of objects to be prioritized in this analysis is less obvious to be determined or even not certain prior to carrying out the AHP. This type of analysis is actually aimed at testing the inconsistency between the right and left eigenvector approaches by modelling real life situations. Generally, in ranking a lower level with respect to its immediate upper level, there will not be absolute dominance of one object over others. Based on this type of analysis, the extent of right and left eigenvector inconsistency can be checked particularly for the overall ranking result. The following steps will illustrate this.

Step 1 - Rank Level 2 (Criteria) with respect to Level 1 (Goal)

Under "less obvious order of objects", this reciprocal matrix could be generated randomly.

Step 2 - Rank Level 3 (Alternatives) with respect to Level 2 (Criteria)

Using Criterion 1 as objective, the ranking order of the three alternatives is, for example, Alternative C, then Alternative B and then Alternative A.

Using Criterion 2 as objective, the ranking of the three alternatives is, for example, Alternative A, then Alternative B and then Alternative C.

Using Criterion 3 as objective, the ranking of the three alternatives is, for example, Alternative C, then Alternative A and then Alternative B.

Using Criterion 4 as objective, the ranking of the three alternatives is, for example, Alternative B, then Alternative A and then Alternative C.

2.2.2 Results Obtained by Tung and Tang

Based on the 294 reciprocal matrices formed by the 42 test models (21 models for pre-arranged order of objects and 21 models for less obvious order of objects), the right and left eigenvector inconsistency was tested. Readers should refer to Tung and Tang (1998) for the explanation on the details how 294 (i.e. 147 + 147) matrices were formed.

For the 147 reciprocal matrices of pre-arranged order of objects, there were 6 and 18 sets of rank reversals between the right and left eigenvector approaches for the SAHP and MAHP respectively. From the SAHP using the right and left eigenvector approaches, only 5 and 4 sets of results could not successfully yield the correct pre-determined order of objects respectively. Similarly, from the MAHP using right and left eigenvector approaches, 17 and 15 sets of results could not successfully yield the correct pre-determined order of objects respectively. Furthermore, both SAHP and MAHP, using right and left eigenvector approaches, can successfully yield the correct order of the alternatives in the overall ranking result as the presumed ones for all the 21 models even though some inconsistencies and rank reversals have occurred in some individual reciprocal matrices.

For the 147 reciprocal matrices of less obvious order of objects, there were 17 and 15 sets of rank reversals between the two eigenvector approaches for the SAHP and MAHP respectively. Moreover, 6 out of 21 models for the SAHP can yield consistent overall ranking results using the right and left eigenvector approaches, and 9 out of 21 models for the MAHP can yield the same.

Based on the above results, it was found that, overall speaking, MAHP is **no better** than SAHP on the issue of rank reversals.

2.3   Geometric Mean for Deriving Priority Vectors

Barzilai (1997) suggested that in order to eliminate the inconsistent results of the right and left eigenvector approaches, **geometric mean**, instead of eigenvector, should be used to derive a priority vector. Let us look at Example 2.4.

Example 2.4

Find the priority vector of the reciprocal matrix **A** in Example 2.1 and that of $\mathbf{A}^\mathrm{T}$ by geometric mean method.

Solution

$$A = \begin{bmatrix} 1 & 1 & 3 & 9 & 9 \\ 1 & 1 & 5 & 8 & 5 \\ 1/3 & 1/5 & 1 & 9 & 5 \\ 1/9 & 1/8 & 1/9 & 1 & 1 \\ 1/9 & 1/5 & 1/5 & 1 & 1 \end{bmatrix}$$

$$\text{Priority vector} = \begin{bmatrix} (1 \times 1 \times 3 \times 9 \times 9)^{1/5} \\ (1 \times 1 \times 5 \times 8 \times 5)^{1/5} \\ (1/3 \times 1/5 \times 1 \times 9 \times 5)^{1/5} \\ (1/9 \times 1/8 \times 1/9 \times 1 \times 1)^{1/5} \\ (1/9 \times 1/5 \times 1/5 \times 1 \times 1)^{1/5} \end{bmatrix} = \begin{bmatrix} 3.000 \\ 2.885 \\ 1.245 \\ 0.274 \\ 0.339 \end{bmatrix} \begin{matrix} - 1^{st} \\ - 2^{nd} \\ - 3^{rd} \\ - 5^{th} \\ - 4^{th} \end{matrix}$$

Readers can see that the ranking of the elements in this example is the same as the priority vector obtained from the left eigenvector approach in Example 2.1. It should be noted that the priority vector in this example is **normalized multiplicatively**, that is, $\Pi x_i = 1$, and no longer **additively** ($\Sigma x_i = 1$). The latter is applicable only in eigenvector ranking.

$$A^T = \begin{bmatrix} 1 & 1 & 1/3 & 1/9 & 1/9 \\ 1 & 1 & 1/5 & 1/8 & 1/5 \\ 3 & 5 & 1 & 1/9 & 1/5 \\ 9 & 8 & 9 & 1 & 1 \\ 9 & 5 & 5 & 1 & 1 \end{bmatrix}$$

$$\text{Priority vector} = \begin{bmatrix} (1 \times 1 \times 1/3 \times 1/9 \times 1/9)^{1/5} \\ (1 \times 1 \times 1/5 \times 1/8 \times 1/5)^{1/5} \\ (3 \times 5 \times 1 \times 1/9 \times 1/5)^{1/5} \\ (9 \times 8 \times 9 \times 1 \times 1)^{1/5} \\ (9 \times 5 \times 5 \times 1 \times 1)^{1/5} \end{bmatrix} = \begin{bmatrix} 0.333 \\ 0.346 \\ 0.803 \\ 3.650 \\ 2.954 \end{bmatrix} \begin{matrix} - 1^{st} \\ - 2^{nd} \\ - 3^{rd} \\ - 5^{th} \\ - 4^{th} \end{matrix}$$

Readers can see that the ranking of the priority vector of A^T is the same as that of A. No rank reversal is observed when the geometric mean method is used. Let us now see another property of the geometric mean solutions as follows:

Let $g(A)$ = the priority vector of the reciprocal matrix A by geometric mean method

$$g(A) = \begin{bmatrix} 3.000 \\ 2.885 \\ 1.245 \\ 0.274 \\ 0.339 \end{bmatrix} \quad \text{as obtained in the first part of Example 2.4}$$

$$g(A^T) = \begin{bmatrix} 0.333 \\ 0.346 \\ 0.803 \\ 3.650 \\ 2.954 \end{bmatrix} \quad \text{as obtained in the second part of Example 2.4}$$

$$1 \div g(A^T) = [\, g(A^T)\,]^{-1} = \begin{bmatrix} (0.333)^{-1} \\ (0.346)^{-1} \\ (0.803)^{-1} \\ (3.650)^{-1} \\ (2.954)^{-1} \end{bmatrix} = \begin{bmatrix} 3.000 \\ 2.885 \\ 1.245 \\ 0.274 \\ 0.339 \end{bmatrix}$$

Therefore, $g(A) = 1 \div g(A^T)$

Barzilai in his paper of 1997 concluded that geometric mean is the only method for deriving weights from multiplicative pairwise comparisons which satisfies fundamental consistency requirement. It is immune to scale-inversion or left-right rank reversal. He thinks that it is the only solution which preserves the strong algebraic structure of the problem.

2.4   Simplified AHP

In Sections 1.3 through 1.5 of Chapter 1, the method for evaluating multi-level hierarchy problems is discussed. It can be observed that all the vectors used in the process of evaluation are normalized additively. In this section, a simplified approach (Tung, 1997) is proposed to solve multi-level hierarchy problems using such normalization property.

Suppose one have to compare four objects with respect to a criterion and they are Objects A, B, C and D. Now, one can arrange the four objects in ascending or descending order with respect to the criterion. The order is assumed to be, say: Object A is better than Object B, Object B is better than Object C and Object C is better than Object D.

Firstly, let the rating of Object D (the smallest rating) be x and suppose the relative importance of Object C, B and A are 2x, 5x and 7x respectively. Of course, the relative ratings of the objects are solely on the subjective judgements of the decision maker.

Since the ratings of the four objects are summed to unity (i.e. normalized), the above expression can be written in a simple equation as follows:

$$\text{Object:} \quad A \qquad B \qquad C \qquad D$$
$$7x \; + \; 5x \; + \; 2x \; + \; x \; = 1$$

After solving the above equation, x is found to be 0.067 and the relative ratings of the four objects can be worked out as:

Object	Rating	Ranking
A	0.467	1
B	0.333	2
C	0.133	3
D	0.067	4

For full hierarchical analysis, all cluster analyses can be handled in the same way and all the priority vectors can be evaluated as shown above.

A worked example is to be given here to explain clearly how this simplified AHP method can be applied to full hierarchical analysis.

Example 2.5

A three level hierarchy with four criteria (m = 4) at level 2 and 3 alternatives (n = 3) at level 3 is shown as follows:

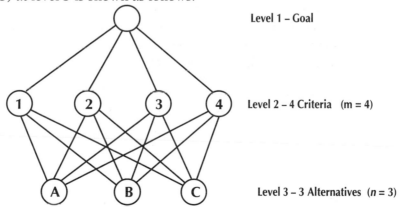

Level 1 – Goal

Level 2 – 4 Criteria (m = 4)

Level 3 – 3 Alternatives (n = 3)

Solution

To analyze the problem using the simplified AHP, the following steps are typically carried out:

Step 1 - Rank Level 2 (Criteria) with respect to Level 1 (Goal)

Instead of carrying out pairwise comparisons, the cluster analysis can be handled as follows. Suppose, with respect to Level 1 (the Goal), the four criteria can be arranged in descending order as : Criterion 1 is better than Criterion 2, Criterion 2 is better than Criterion 3 and Criterion 3 is better than Criterion 4.

Let the rating of Criterion 4 be x_1 and the relative ratings of Criteria 3, 2 and 1 be $2x_1$, $3x_1$ and $4x_1$ respectively. Then the following equation can be obtained.

$$\text{Criteria:} \quad 1 \qquad 2 \qquad 3 \qquad 4$$
$$4x_1 \; + \; 3x_1 \; + \; 2x_1 \; + \; x_1 \; = \; 1$$

By solving the above equation, the relative ratings of the four criteria can be worked out and presented in Matrix X as:

$$X = \begin{bmatrix} 0.400 \\ 0.300 \\ 0.200 \\ 0.100 \end{bmatrix} \begin{matrix} \text{Criterion 1} \\ \text{Criterion 2} \\ \text{Criterion 3} \\ \text{Criterion 4} \end{matrix}$$

Step 2 - Rank Level 3 (Alternatives) with respect to Level 2 (Criteria)

(i) Using Criterion 1 as objective

With respect to Criterion 1, it is assumed that the order of the three alternatives is A, then B, and then C. Let the rating of Alternative C be y_1 and the relative ratings of Alternatives B and A be $3y_1$ and $5y_1$ respectively. The following equation can be obtained.

$$\text{Alternatives:} \quad A \qquad B \qquad C$$
$$5y_1 \; + \; 3y_1 \; + \; y_1 \; = \; 1$$

By solving the above equation, the relative ratings of the three alternatives can be worked out and presented in Matrix Y_1 as:

$$Y_1 = \begin{bmatrix} 0.556 \\ 0.333 \\ 0.111 \end{bmatrix} \begin{matrix} \text{Alternative A} \\ \text{Alternative B} \\ \text{Alternative C} \end{matrix}$$

(ii) Using Criterion 2 as objective

With respect to Criterion 2, it is assumed that the order of the three alternatives is C, then A, and then B. Let the rating of Alternative B be y_2 and the relative ratings of Alternatives A and C be $4y_2$ and $6y_2$ respectively. The following equation can be obtained.

$$\begin{matrix} \text{Alternatives:} & A & B & C \\ & 4y_2 + & y_2 + & 6y_2 = 1 \end{matrix}$$

By solving the above equation, the relative ratings of the three alternatives can be worked out and presented in Matrix Y_2 as:

$$Y_2 = \begin{bmatrix} 0.364 \\ 0.091 \\ 0.545 \end{bmatrix} \begin{matrix} \text{Alternative A} \\ \text{Alternative B} \\ \text{Alternative C} \end{matrix}$$

(iii) Using Criterion 3 as objective

With respect to Criterion 3, it is assumed that the order of the three alternatives is B, then C, and then A. Let the rating of Alternative A be y_3 and the relative ratings of Alternatives C and B be $2y_3$ and $3y_3$ respectively. The following equation can be obtained.

$$\begin{matrix} \text{Alternatives:} & A & B & C \\ & y_3 + & 3y_3 + & 2y_3 = 1 \end{matrix}$$

By solving the above equation, the relative ratings of the three alternatives can be worked out and presented in Matrix Y_3 as:

$$Y_3 = \begin{bmatrix} 0.167 \\ 0.500 \\ 0.333 \end{bmatrix} \begin{matrix} \text{Alternative A} \\ \text{Alternative B} \\ \text{Alternative C} \end{matrix}$$

(iv) Using Criterion 4 as objective

With respect to Criterion 4, it is assumed that the order of the three alternatives is C, then B, and then A. Let the rating of Alternative A be y_4 and the relative ratings of Alternatives B and C be $7y_4$ and $8y_4$ respectively. The following equation can be obtained.

$$\text{Alternatives:} \quad A \qquad B \qquad C$$
$$y_4 \quad + \quad 7y_4 \quad + \quad 8y_4 \; = \; 1$$

By solving the above equation, the relative ratings of the three alternatives can be worked out and presented in Matrix Y_4 as:

$$Y_4 = \begin{bmatrix} 0.062 \\ 0.438 \\ 0.500 \end{bmatrix} \begin{matrix} \text{Alternative A} \\ \text{Alternative B} \\ \text{Alternative C} \end{matrix}$$

Step 3 - Formulate the composite matrix C such that $C = [Y_1, Y_2, Y_3, Y_4]$, that is,

$$C = \begin{bmatrix} 0.556 & 0.364 & 0.167 & 0.062 \\ 0.333 & 0.091 & 0.500 & 0.438 \\ 0.111 & 0.545 & 0.333 & 0.500 \end{bmatrix}$$

Let R = the matrix representing the overall ranking of the four alternatives

$$R = C \times X = \begin{bmatrix} 0.556 & 0.364 & 0.167 & 0.062 \\ 0.333 & 0.091 & 0.500 & 0.438 \\ 0.111 & 0.545 & 0.333 & 0.500 \end{bmatrix} \begin{bmatrix} 0.400 \\ 0.300 \\ 0.200 \\ 0.100 \end{bmatrix}$$

$$\qquad\qquad\qquad\qquad\qquad\qquad\qquad\qquad \text{Ranking}$$
$$= \begin{bmatrix} 0.371 \\ 0.304 \\ 0.325 \end{bmatrix} \begin{matrix} \text{Alternative A} \\ \text{Alternative B} \\ \text{Alternative C} \end{matrix} \quad \begin{matrix} 1 \\ 3 \\ 2 \end{matrix}$$

Hence, from the overall ranking matrix R, Alternative A should be selected as the best alternative since it has the highest value.

Exercise Questions

Question 1

Reference is made to the paper written by V. Belton and T. Gear entitled "On a short-coming of Saaty's method of analytic hierarchies" in *Omega*, 1983 (Vol. 11, No. 3, pp.228-230). What are the examples on rank reversal given by Belton and Gear?

Question 2

Reference is made to the paper written by J. Barzilai and B. Golany entitled "AHP rank reversal, normalization and aggregation rules" in INFOR, 1994 (Vol. 32, No. 2, pp.57-64). How did Barzilai and Golany use geometric mean approach to resolve the rank reversal problem of Belton and Gear's? Compare Barzilai and Golany's result with those of SAHP and MAHP.

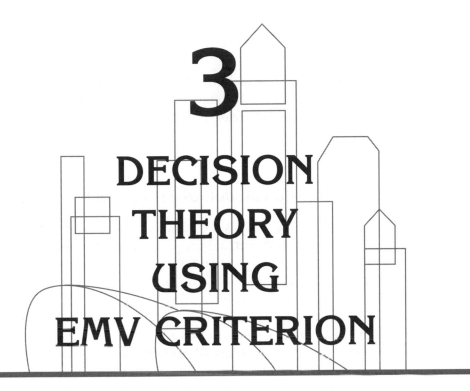

3
DECISION
THEORY
USING
EMV CRITERION

3.1　Decision Analysis

In day-to-day work, construction managers may face problems which involve probability, that is, for any one particular action that an engineer takes, there may be several probable outcomes.

Let us consider a simple example. A contractor has to decide whether to hire concrete pumping equipment in order to complete a foundation work tomorrow. If he does not hire the equipment, work will be delayed and he will suffer a loss of $10,000. If he hires the equipment, he anticipates two possible outcomes, depending on the weather:

　a.　　If the weather is fine, the concrete pumping equipment will be fully utilized. The concrete will be completed and he will gain a profit of $20,000 after deducting the rent for the equipment.

　b.　　If there is rain, the equipment cannot function and the work will be delayed. The contractor will have to pay $10,000 rent and he will also suffer a loss of $10,000 for the delay of work.

In the above example, the two outcomes involve probability. But it is possible for the contractor to estimate, by the guesswork or with the help of historical rainfall records provided by the Meteorological Observatory, the probability of each outcome.

Having determined the probability of the two possible outcomes, he can make a decision whether or not to hire the equipment. This chapter describes a mathematical method called **decision analysis** which helps the contractor reach a decision.

3.2   Building a Decision Model: Decision Tree Formation

The first step in decision analysis is to identify the various **actions** which can be taken at a **decision point** and the probable **outcomes** of each action taken. In the example given in Section 1, the contractor has arrived at a decision point where he can take either of the two actions (that is, hire or not hire the equipment). There is only one outcome of the first action (there is delay of work and he suffer a loss). The second action has two probable outcomes (either the weather is fine or there is rain). Other information needed for decision analysis is the **probability of occurrence** of each outcome, together with its outcome value (a measure of the profit or loss associated with the outcome). In this example, the probability of occurrence of each outcome has not been further investigated, but it can be known after obtaining past rainfall records from the Meteorological Observatory. The outcome value is the profit depending on the weather conditions (for example, the outcome value if the weather is fine is $20,000 and the value if it rains is -$20,000).

The data is usually represented in a **decision tree**, which forms the **decision model** of the problem. Another example is given below to demonstrate how a decision tree is drawn.

Example 3.1

A contractor succeeds in bidding for a contract at a tender price of $20 million. He estimates that there are three possible outcomes in completing the contract:

i   the actual construction cost is $15 million (i.e. profit $5 million) with a probability of 0.2;

ii.   the actual construction cost is $17 million (i.e. profit $3 million) with a probability of 0.65; or

iii. the actual construction cost is $19 million (i.e. profit of $1 million) with a probability of 0.15.

Our task is to express the above data in the form of a decision tree for the contractor.

Solution

The decision tree is shown below in Fig. 3.1:

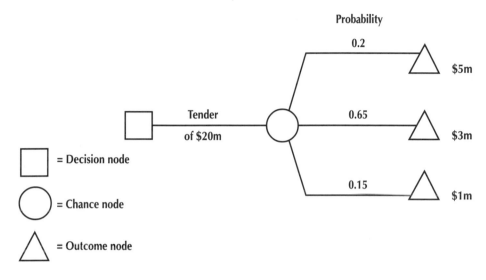

Fig. 3.1 Fxample of a very simple decision tree

A **decision node** is a branching point in a decision tree where alternative **actions** can be taken. In this example, the contractor offers only one tender (one action). Hence, there is only one branch from this node (see Fig. 3.1).

Branching out from an action is a **chance node**. It is a branching point in a decision tree where various possible outcomes occur. In this example, there are three possible outcomes of the chance node. The probability of each outcome is written on the branch line.

Corresponding to each branch from a chance node is an **outcome node**. In Fig. 3.1, the possible outcome values, or profits of each outcome ($5 million, $3 million and $1 million), have been written next to each outcome node.

Let us now consider a more realistic example.

Example 3.2

A contractor is performing some work in a coastal area which is subject to destructive typhoon conditions. During the course of the construction work he has to provide storage for his plant for a week. The contractor can either leave the plant on site or he can move it to a typhoon-safe storage place.

If he keeps the plant idle on site, he can either build a protective shelter at a cost of $30,000 which will protect the plant against minor (but not major) typhoons, or he can leave the plant unprotected on site with no cost incurred. However, he then risks losing the plant if a typhoon, both major or minor, occurs.

On the other hand, he can take the plant off the site and move it to a safe place, store it there for a week, and then move it back to the site at a total cost of $42,500.

The plant, which cost $400,000, will be destroyed by either a minor or major typhoon if it is left unprotected on site. A major typhoon but not a minor typhoon will also destroy it if a protective shelter is built on site. Typhoons cannot damage the plant if it is stored in the safe place away from the site. The probabilities of being hit by a major or a minor typhoon during the week, as estimated from the past records of the Meteorological Observatory, are 0.01 and 0.09 respectively.

The task is to draw a decision tree for the contractor.

Solution

(i) The decision tree if the plant is left unprotected is given below.

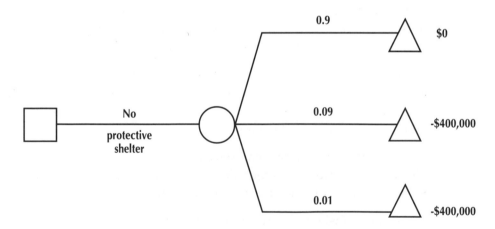

Fig. 3.2 Decision tree for unprotected plant

(ii) The decision tree if the plant is left on site protected by a shelter is given below:

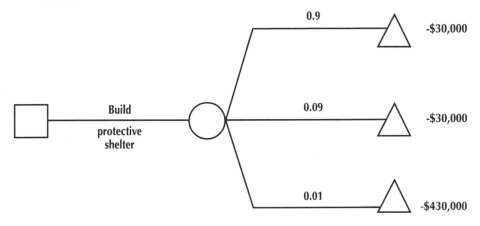

Fig. 3.3a Decision tree for plant protected by a shelter

It can also be presented as shown in Fig. 3.3b. In fact it is a **better** presentation than Fig. 3.3a.

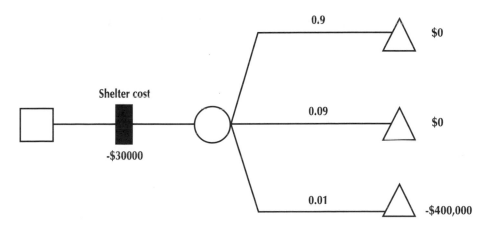

Fig. 3.3b Decision tree for plant protected by a shelter
(using alternative presentation)

(iii) The whole problem can be presented by the following decision tree:

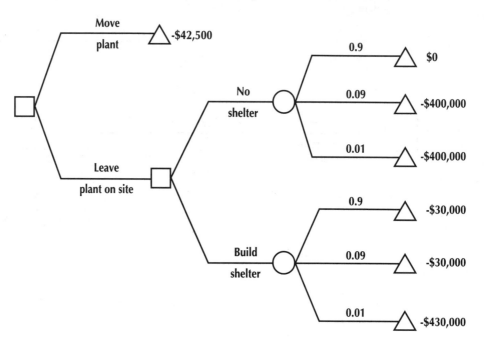

Fig. 3.4a Decision tree representing the whole problem

It can also be represented as shown in Fig. 3.4b. As mentioned before, it is a **better** form of presentation than Fig. 3.4a. In certain cases, such as decision trees for finding the value of information (see Chapter 5), this form of presentation **must** be used.

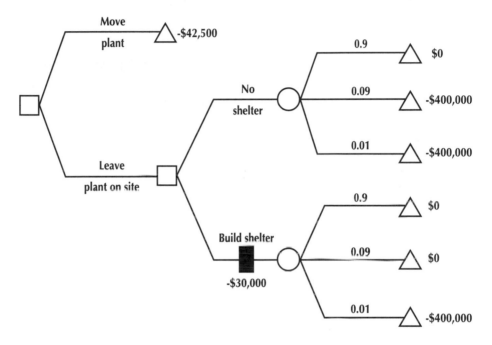

Fig. 3.4b Decision tree representing the whole problem
(using alternative presentation)

3.3   Expected Monetary Value (EMV)

Before going on to discuss the procedure for finding the optimal decision, we have to know what an **expected monetary value (EMV)** attached to a particular action is.

The EMV of the outcome branching from a chance node is defined as the sum of the products of the outcome values and their respective probabilities. Mathematically, it can be written as follows:

$$\text{EMV} = \sum_{i=1}^{n} p_i x_i$$

where n = number of outcomes branching from a chance node

x_i = value of outcome i, and

p_i = probability of occurrence of outcome i.

The EMV of the outcomes for Example 3.1 in Section 3.2 is therefore calculated as 0.2 x 5,000,000 + 0.65 x 3,000,000 + 0.15 x 1,000,000 = 3,100,000.

The EMV is $3,100,000. This is the **expected profit** of the contract. It must be noted, however, that the profit arising from this contract may not be exactly $3,100,000. This profit is only an average figure. This means that if a considerable number of similar contracts were to be taken, then the profit for such contracts would be $3,100,000 on average. It is unlikely that this particular contract will bring a profit of exactly $3,100,000 if it is done just once.

The EMV definition is in fact obtained from the definition of average (or mean, or expected) value. In statistics, the definition of an average value is given by the following expression:

$$\text{Average} = \frac{\sum_{i=1}^{i=n} f_i x_i}{\sum_{i=1}^{i=n} f_i}$$

where x_i = value of outcome i (i=1, ...n), and

f_i = frequency of occurrence of outcome i.

The EMV of a set of outcomes is actually the average of the outcome values. The proof is given below:

Average of the outcome values $= (f_1 x_1 + f_2 x_2 + ... + f_n x_n) / \sum f_i$

$$= (f_1/\sum f_i)x_1 + (f_2/\sum f_i)x_2 + ... + (f_n/\sum f_i)x_n$$

$$= p_1 x_1 + p_2 x_2 + ... + p_n x_n$$

(where p_i = probability of event i)

Therefore, average $= \sum_{i=1}^{i=n} p_i x_i$

$$= \text{EMV}$$

3.4   Using EMV to Solve a Decision Problem

Section 3.2 shows how to represent a decision problem in the form of a decision tree, and Section 3.3 explains the EMV criterion. The next step is to solve a decision problem.

Let us use the decision model in Example 3.2 as an example. The problem may be solved using a number of steps.

Step 1

Set up the decision tree for the problem.

The model is reproduced below for easy reference. For convenience, the decision nodes are labeled A and B and the chance nodes are numbered 1 and 2.

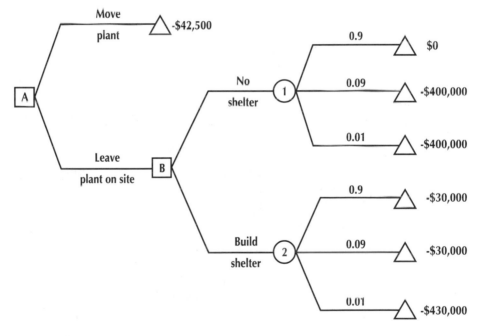

Fig. 3.5 Decision tree representing the problem in Example 3.2

Step 2

Find the EMVs at the various chance nodes and enter the values onto the decision tree.

The EMV at chance node 1 (EMV_1) is -$40,000 and the EMV at chance node 2 (EMV_2) is -$34,000. The calculations are left to the readers

Step 3

At decision node B, compare the EMVs and select the one with a higher value (i.e. maximum profit or minimum loss). In this case, EMV_2 is selected as it represents the minimum loss. Write X on the route which has not been selected.

At decision node A, compare the EMVs of the two routes which are left uncompared and select the one with the smaller loss. The EMVs here are -$34,000 ($EMV_2$) and -$42,500 and so EMV_2 is selected and an X is put on the unselected route.

Choose the path without an X on the route. This path represents the optimal decision.

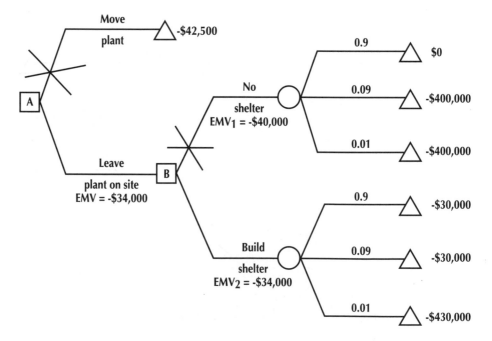

Fig. 3.6 Finding the optimal path

We can now reach a decision. The plant should not be moved, but a shelter should be built to protect it from minor typhoons. Such a decision has an expected monetary value of -$34,000. The negative sign indicates that it is a loss rather than a profit.

The same answer can be obtained if Fig 3.4b is used. This exercise is left to the readers. It is here reiterated that Fig.3.4b is a better form and should be encouraged to use, although the above example has not used Fig 3.4b to illustrate the calculation.

3.5   Summary

The following is a summary of the steps for making optimal decision based on the EMV criterion.

1. Represent a decision problem by a decision tree. The tree is drawn after alternative actions are identified and the values of the outcomes and their respective probabilities are known.
2. Use the EMV criterion as the basis for decision making. Calculate the EMVs at the chance nodes. Start with the chance nodes nearest to the tips of the tree (at the most right-hand side of the tree).
3. Beginning with the tip of the decision tree, compare the EMVs of the alternative actions at each decision node. Select the alternative action with the highest EMV (i.e. the maximum expected profit or minimum expected loss) and assign this EMV to that decision node.
4. Proceed node by node from right to left. Put an X on the routes which have not been selected.
5. The overall decision is given by the route (from left to right) with no X signs.

Exercise Questions

Question 1

A small building contracting firm is currently concentrating on relatively small jobs and is considering upgrading the size of its contracts. If the conpany works on small projects, the profit for the coming year is estimated to be $2,000,000. However, there are two considerably large contracts being offered to the company. They are:

(i)   a school building contract of about one year's duration (Contract X); and

(ii) a supermarket building contract also of about one year's duration (Contract Y).

Tenders for Contract X must be submitted at the end of the month and the successful tenderer will be notified in the second week of the following month. Tenders for Contract Y must be submitted at the middle of next month and the successful tenderer will be announced at the end of that month. So, if the contractor is unsuccessful in bidding for Contract X, he will still have time to bid for Contract Y.

Either of these two contracts, if won, will commence within a month of notification and will absorb all the contractors money available to him. This means that the contractor cannot tender for Contract Y if he succeeds in bidding for Contract X.

The contractor has assessed his probability of success in bidding for the two contracts. He can enter either a high tender price or a low one for each contract. Of course, when he bid a high price tender, his chance of getting the contract is lower, and vice versa. The probability of success and profit attached to each outcome are shown in the table below:

	Contract X		Contract Y	
	Profit	Probability of successful bidding	Profit	Probability of successful bidding
High tender price	$4,000,000	0.4	$5,000,000	0.35
Low tender price	$3,000,000	0.7	$3,500,000	0.8

What should be the company's strategy on bidding these two contracts?

Question 2

A construction firm has a plant department in its head office. The department, in addition to allocating plant to various sites where construction works are being undertaken by the firm, rents out the surplus plant to other contracting firms. There are four items of surplus plant. The probabilities of demand for plant by outside firms on any day, according to past records, are given below:

Number of items of plant demanded	Probability
0	0.1
1	0.2
2	0.25
3	0.3
4	0.15

The hiring charge is $3,000 per day for each plant. The average maintenance (i.e. overhead) cost of each plant to the firm is $800 a day, whether or not the plant is hired out. What is the daily expected profit for the plant hiring business?

Suppose the manager of the plant department is about to review its business performance in order to stock the optimal number of plant for hire. Determine this optimal number of surplus plant which the department manager should have in stock so that he can make the greatest profit.

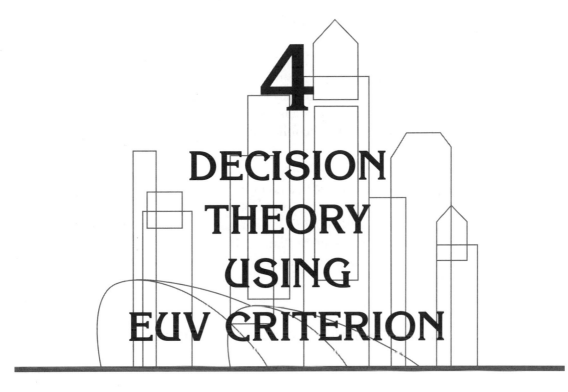

4
DECISION THEORY USING EUV CRITERION

4.1 The Drawbacks of the EUV Criterion

In Section 3.3 of Chapter 3, it was explained that the EMV criterion in decision analysis is based on the law of averages. If the decision maker can afford to take a long-term view and accept temporary losses and gains, such a decision may pay off in the long run.

The EMV criterion provides an objective measure of money value. However, it does not take into account people's subjective views towards different amounts of money and the degree of risk people are willing to take. It assumes that different people have equal satisfaction when they gain a certain amount of money and that they have equal dissatisfaction when they lose the same amount of money. This assumption is not true in practice. A poor man who suddenly obtains an unexpected sum of $10,000 would probably be much happier than a millionaire who suddenly receives that amount. A poor man who has saved $10,000 would also be more unhappy if he loses all $10,000 he has saved than a millionaire who loses $10,000.

On the other hand, for a same person, the degree of happiness at gaining a fixed amount of money may not be the same as the degree of unhappiness at losing that same amount of money. For example, your amount of unhappiness at losing $10,000 may be greater than your amount of pleasure at gaining $10,000, or vice versa.

In short, the EMV criterion fails to take into account of different levels of importance of a given sum of money for different people. This is particularly true in cases where the decision maker is in a position of restricted financial resources. Let us consider the following example.

Example 4.1

A retired man receives a pension amounting to $200,000. He wishes to make good use of it by investing it in a profitable and secure business. Investment Plan I provides him with a 0.7 chance of losing the $200,000, but a 0.3 of gaining $1 million. Investment Plan II gives him a 0.6 chance of gaining $150,000 and a 0.4 chance of gaining $160,000. Which investment should he choose?

Solution

In order to solve this problem, let us first represent it in the form of a decision tree:

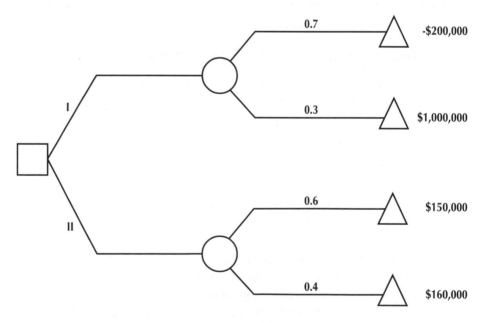

Fig. 4.1 Decision tree representing the problem in Example 4.1

If the man employs the EMV criterion to make a decision, he should choose investment Plan I, since the EMV of Plan I is $160,000, whereas that of Plan II is only $154,000 (check these values yourself).

Let us now analyze the problem using just our common sense. The retired old man's total assets are $200,000. It would be disastrous if he were to lose it. Investment Plan I has 0.7 chance of losing all his money, although he may gain $1 million at a 0.3 chance. It would be much safer, therefore, for him to choose investment Plan II, which could give him, in addition to financial security, a profit of either $150,000 (at a 0.6 chance) or $160,000 (at a 0.4 chance).

In this example, the decision based on the EMV criterion is therefore unsatisfactory for the decision maker. This is not because the expected profit of $160,000 is less attractive than $154,000, but simply because the thought of losing $200,000 is completely unacceptable to the retired man.

This drawback of the EMV theory can be overcome by using what is known as the **utility value theory**. This theory is designed to take into consideration the subjective or personal view of the importance of a gain or loss to a decision maker.

4.2   The Utility Value

The example above shows that the decision maker would find it absolutely unacceptable to loss $200,000, even though a gain of $1 million would give him much pleasure. He is not just considering the expected monetary gain but is thinking of the **utility value** of gaining and losing the said amounts of money. The utility value refers to the pleasure (or displeasure) one would get from gaining (or losing) a specific amount of money.

A **utility curve** can be used to illustrate the relationship between the money gained (or lost) by a person and the degree of pleasure (or displeasure) which will be gained (or lost).

4.3   Developing a Utility Curve

The monetary values in a problem must be transformed into utility values before they can be used to make a decision. This transformation is possible if the utility curve for the decision maker is known.

Let us go back to the problem in Example 4.1. For convenience, the decision tree for the problem is reproduced below (Fig. 4.2).

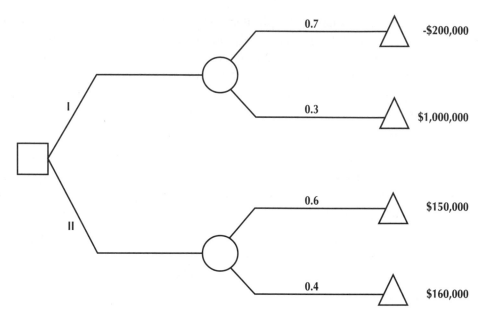

Fig. 4.2 Decision tree for the problem

There are three steps in the development of a utility curve. They are explained below.

Step 1

Identify the largest and the smallest monetary values in the decision tree.

In this problem, the largest monetary value is $1 million and the smallest monetary value is -$200,000.

Step 2

Assign arbitrary utility values to the largest and smallest monetary values as shown in Table 4.1.

	Monetary Value ($)	Utility value (utiles)
Largest	1,000,000	100
Smallest	-200,000	0

Table 4.1 Assigning utility values to the largest and smallest monetary values

Now we have two points (see Fig. 4.3) on the utility curve simply by assigning the two utility values to the two monetary values.

Note that the unit for monetary value is **dollar**, while the unit for the utility value is **utile**.

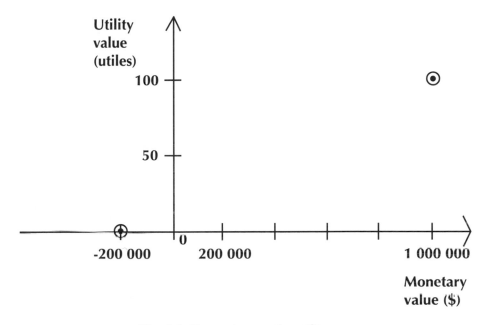

Fig. 4.3 Two points on the utility curve

Step 3

Generate another utility value for another monetary value by using a hypothetical decision problem. (This is the most difficult step in the process.)

The basic aim in this step is to establish how much risk the person is willing to take. How willing is he to take the chance in order to get a higher expected monetary value, as opposed to accepting a smaller monetary value with no risk? We can get an answer to this question by assigning an arbitrary amount of money, and asking him what amount of guaranteed payoff he considers to be equivalent to the expected monetary value. Different persons will give different responses, thus creating subjective views which will be reflected by different utility curves. Let us explain this further below.

In order to determine one's response, the decision maker is asked the following question: what amount of money 'x' (which you can obtain for sure) do you consider to be equivalent to a 50:50 gamble on -$200,000 and $1 million? In other words, how much would you want to receive, rather than take the chance of a 50:50 gamble on winning $1 million or losing $200,000?

The answer to this question enables us to find the utility value for a particular monetary value.

To find the answer to this question is the same as finding 'x' such that the decision maker considers the two alternatives in Fig. 4.4 to be equivalent in terms of utility value.

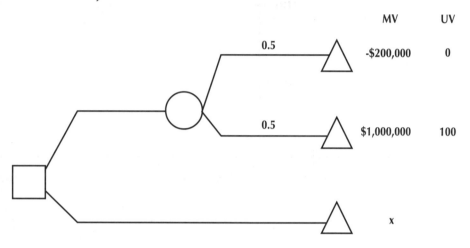

Fig. 4.4 Decision tree giving monetary values and utility values

In this example, the **expected utility value** of the first alternative (the risky alternative) is (100 x 0.5) + (0 x 0.5) = 50, since 100 and 0 are the utility values of $1 million and -$200,000 respectively.

If the decision maker (let us call him A) does not like risks, his answer to the above question might be, for example, $300,000. We can then say that A is a **risk averter** because the EMV of the risk is $400,000, but, for security reasons, he only asks for $300,000. In the case of decision maker A, x = $300,000 is considered to have an equivalent utility value of 50. This gives another point to plot on A's utility curve (i.e. $300,000, 50) — see Fig. 4.5.

The answer of an even more conservative decision maker (B), who is even less willing to take chances than A, may be x = $150,000. For decision maker B, the utility value of $150,000 is 50.

A **gambler** is more willing to take chances, and will only accept an amount which is larger than the expected monetary value $400,000 by taking the chance. In this example, if the decision maker (C) is a gambler, he will consider, say, x = $500,000 is the amount that will make him satisfied by selling the chance of winning $1,000,000 (although also an equal chance of losing $200,000). In the case of decision maker C, the utility value of $500,000 is 50.

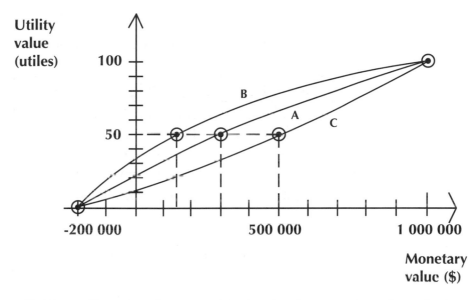

Fig. 4.5 Utility curves for A, B and C, showing the respective values of x for the three decision makers

Step 3 can be repeated to generate more points on the curve. (It should be noted, however, that the risk ratio used in this hypothetical problem (50: 50) was chosen only for convenience). Another question for decision maker A might have been: what amount of money (which you can obtain for sure) do you consider to be equivalent to a 40:60 gamble on -$200,000 and $1,000,000? (See exercise question 2 at the end of this chapter.)

Or the next question might be: what amount of money (which you can obtain for sure) do you consider to be equivalent to a 30:70 gamble on -$200,000 and $300,000?

A smooth curve obtained by joining those points obtained in steps 2 and 3 is the utility curve for the decision maker. Curve A of Fig. 4.5 is the utility curve of the retired old man in question.

4.4   Making Decision Using the EUV Criterion

Once the utility curve for the retired man (decision maker A) is drawn, the utility value for each monetary value involved in the problem can be found from the curve. The EUV for each decision can then be computed and the optimal decision according to the EUV criteria can be determined.

Let see how the solution can be determined.

The utility values are found from the utility curve (Fig. 4.5) of decision maker A.

Monetary value ($)	Utility value (utiles)
-200,000	0
1,000,000	100
150,000	36
160,000	38

Table 4.2 Utility values for decision maker A

The EUVs of the two investment plans can then be computed.

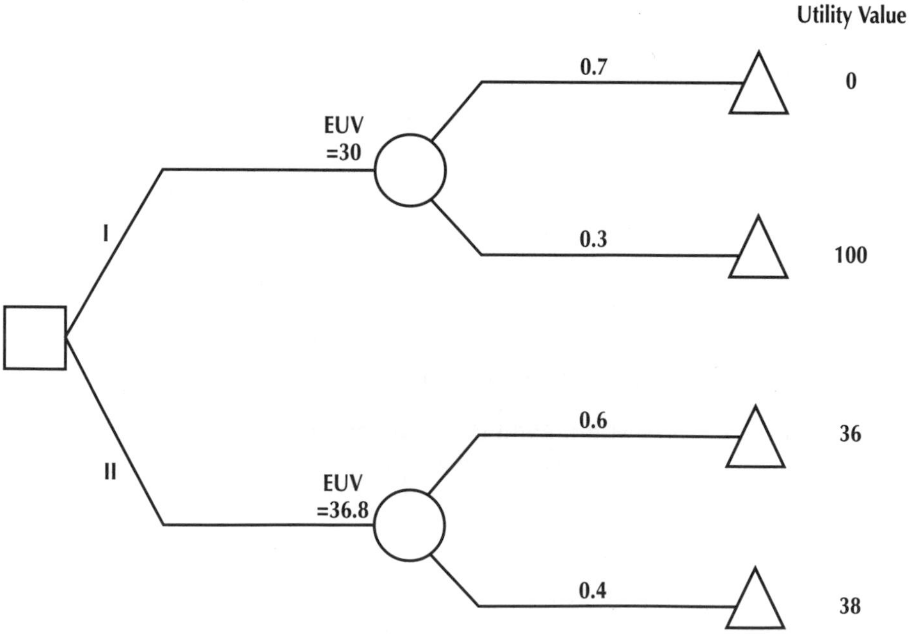

Fig. 4.6 Decision tree represented by utility values

From Fig. 4.6, we can calculate that the EUV of investment Plan I is 30, while that of investment Plan II is 36.8. Hence, according to the utility value criterion, investment Plan II is better than investment Plan I. This result is more consistent with common sense.

4.5   *Further Explanation on the Utility Value Theory*

4.5.1 The straight line joining point 1 and point 2 in Fig. 4.7 represents one possible utility curve (Decision maker D). If this straight line is used as the utility curve, the decision obtained using the EUV criterion will be the same as that obtained by using the EMV criterion.

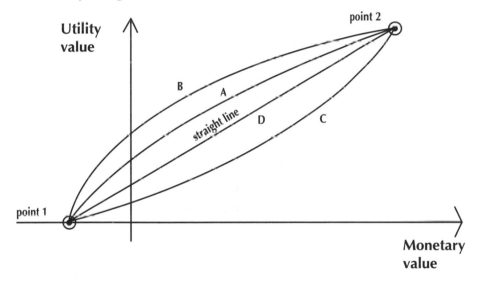

Fig. 4.7 Utility curves

The following is a proof of the statement for Decision maker D.

Suppose that m_{11} and m_{12} are monetary values for action 1 with probabilities p_1 and p_2 respectively; that m_{21} and m_{22} are monetary values for action 2, with probabilities q_1 and q_2 respectively; and that $u_{11}, u_{12}, u_{21}, u_{22}$ are the corresponding utility values,

$$\text{then} \qquad M_1 \quad = \text{EMV of action 1}$$
$$= p_1 m_{11} + p_2 m_{12}$$

$$M_2 \quad = \text{EMV of action 2}$$
$$= q_1 m_{21} + q_2 m_{22}$$

$$U_1 \quad = \text{EUV of action 1}$$
$$= p_1 u_{11} + p_2 u_{12}$$

$$U_2 \quad = \text{EUV of action 2}$$
$$= q_1 u_{21} + q_2 u_{22}$$

Suppose that M (monetary value) and U (utility value) obey a linear relationship, i.e., the case of decision maker D, then

$$U \quad = aM + b$$

where a and b are constants.

Then, $\quad u_{11} \quad = am_{11} + b$
$\qquad u_{12} \quad = am_{12} + b$
$\qquad u_{13} \quad = am_{21} + b$
$\qquad u_{14} \quad = am_{22} + b$

Assume that $M_1 > M_2$

We would like to see if $U_1 > U_2$.

$$\text{Now,} \quad U_1 \quad = p_1 u_{11} + p_2 u_{12}$$
$$= p_1 (am_{11} + b) + p_2 (am_{12} + b)$$
$$= a(p_1 m_{11} + p_2 m_{12}) + b(p_1 + p_2)$$
$$= aM_1 + b$$

Similarly,

$$U_2 \quad = aM_2 + b$$

$$\therefore \quad U_1 \quad = aM_1 + b \quad > \quad aM_2 + b = U_2$$

Thus, the decision obtained by the EUV criterion will be the same as that obtained by the EMV criterion if the relationship between monetary value and utility value is a linear one.

4.5.2 The utility curve A shown in Fig. 4.7 is concave downward and is above the straight line joining the end points. A utility curve of such shape reflects that the investor is a **risk averter**. That is, he avoids risks. This is common for a person who is in a position of restricted financial resources.

4.5.3 A gambler (curve C) is more willing to take risks, although the expected loss is greater (or the expected gain is smaller) when he takes the risk. The utility curve for a gambler would be concave upward and below the straight line (Fig. 4.7).

4.5.4 Arbitrary utility values are assigned to two monetary values in step 2 of Section 4.3. This is because their absolute utility values are not important. Their importance lies only in their relative values: they are used only for comparison at decision nodes.

4.5.5 Step 3, described in Section 4.3, allows a subjective measure of money to take the place of a completely objective measure. This is shown in the utility curve. The shape of a person's utility curve is usually a result of his psychological makeup. Different people will have different utility curves. The same person can have a different utility curve for different types of situations.

4.6   A Further Example

Example 4.2

The contractor in Example 3.2 of chapter 3 is very cautious about the safety of the plant because he is currently in a poor financial position and cannot lose a sum of money larger than $100,000. For him, a 10:90 chance of losing $500,000 to losing nothing is equivalent to losing $100,000 for sure. Find the optimal decision for the contractor.

Solution

The decision tree (based on monetary value) is given in Fig. 4.8.

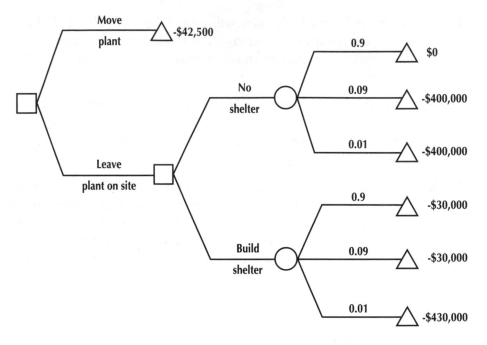

Fig. 4.8 Decision tree for the contractor

All the monetary values in the decision tree have to be transformed into utility values. This is achieved in the following steps.

Step 1

The minimum range of monetary values to be covered in the utility curve is from -$430,000 to $0. There is no harm in taking a larger range (say, from -$500,000 to $0) for plotting the utility curve because the required range of monetary values will be covered.

Step 2

Since utility values can be assigned arbitrarily, we can use values that we think are convenient. Let us assign 0 and 1,000 as utility values for -$500,000 and 0 respectively.

Step 3

The value of 'x' is -$100,000 in the hypothetical problem. The EUV of the 10:90 chance of losing $500,000 to losing nothing is (0.1 x 0) + (0.9 x 1000) = 900. Therefore, the utility value of -$100,000 is 900. (See Fig. 4.9.)

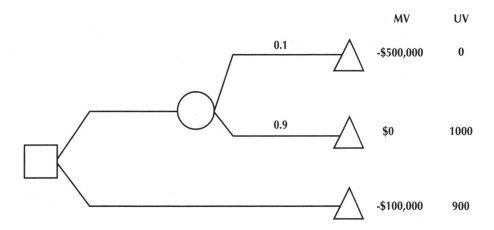

Fig. 4.9 Decision tree of the hypothetical problem reflecting the contractor's risk behaviour

The utility curve can now be sketched (Fig. 4.10).

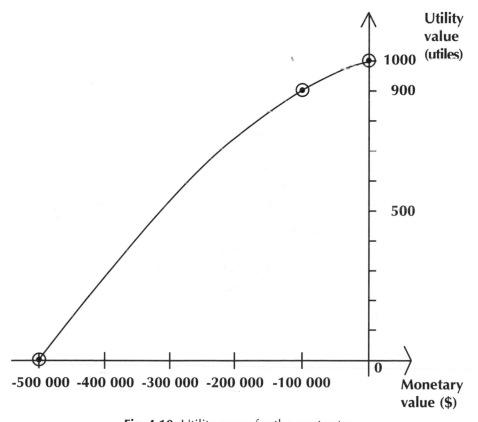

Fig. 4.10 Utility curve for the contractor

Using this utility curve, we can transform all the monetary values in the decision tree to utility values. The EUVs for all actions can then be computed. The EUV criterion can also be used to find the optimal decision. This is done in Fig. 4.11.

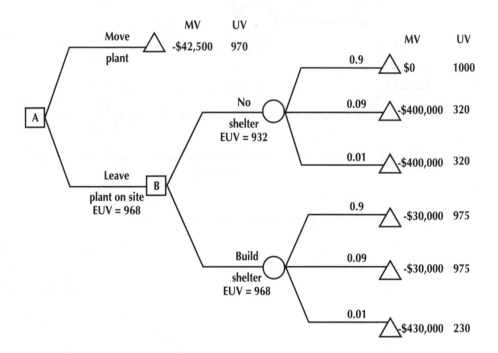

Fig. 4.11 Decision tree with utility values

From the EUV consideration, because the UV 970 is greater than 968, the optimal decision is that the contractor should move the plant away from the site. The worst alternative is to leave the plant unprotected.

Compare this with the result obtained from the EMV criterion. The optimal decision based on the EMV criterion is that the plant should be left on site and a protective shelter built, rather than the plant being moved away from the site (see Section 3.4 of Chapter 3).

Exercise Questions

Question 1

A contractor was successful in bidding for a contract. He is considering buying fire insurance for the site property, which is worth $10,000,000. The premium he will have to pay is $20,000 per year. The probability of having a fire on the site (which he finds out from statistical data) is 0.001 in any one year. Should the contractor buy the insurance,

(a) if the decision is based on the EMV criterion?
(b) if the decision is based on the EUV criterion, with the condition that losing $10,000,000 at a chance of 0.001 is equivalent to losing $30,000 for sure for the contractor?

Question 2

(a) Suppose that the hypothetical decision problem in Fig. 4.4 of Step 3 of Section 4.3 is as follows:

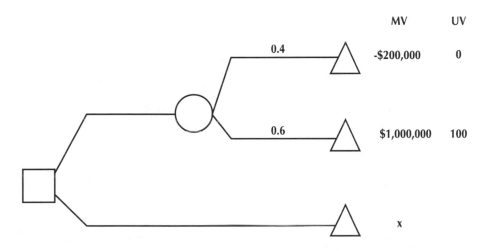

Can we obtain the same curve as shown in Fig. 4.5 as represented by decision maker A?

(b) Suppose that the hypothetical decision problem in Fig. 4.4 of Step 3 of Section 4.3 is as follows:

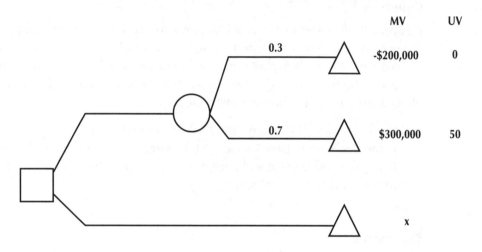

Can we obtain the same curve as shown in Fig. 4.5 as represented by decision maker A?

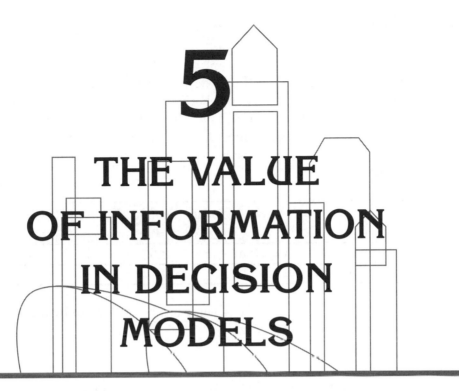

5

THE VALUE
OF INFORMATION
IN DECISION
MODELS

5.1   Problems Involving Conditional Probabilities

Sometimes, a decision problem may involve conditional probabilities. For example, if a construction problem is related to underground soil conditions, engineers can only give predictions. No matter how experienced the engineer is, he cannot predict 100% the true state of the soil conditions underground. Some engineering test, such as seismic test, can, at best, help provide more reliable predictions of the probability of occurrence of various possible states of the variables.

We will see an example of the problems of this sort.

Example 5.1

A contractor is about to drive twenty precast concrete piles to support a structure. It is estimated that there is a 0.5 chance that the depth to firm stratum is 10 m and 0.5 chance that the depth is 15 m. The contractor has to decide whether to order twenty 10m piles or twenty 15 m piles. If 10 m piles are ordered but later found to be too short, he has to spend $150,000 in total to remedy the situation. If 15 m piles are ordered but later found to be too long, he has to cut the piles at a total cost $30,000. The twenty 10 m piles can be bought and driven at a total costs of $250,000; and the twenty 10 m piles can be bought and driven at a total costs of $350,000.

Before committing himself to ordering long or short piles, the contractor has the opportunity to carry out a boring test at a cost of $45,000 to determine the exact depth of the firm stratum. Alternatively, he can carry out a seismic test at a cost of $15,000 to provide more information about the depth of the firm stratum. The seismic test is an imperfect test with the following **sample likelihood**:

	T_1 (true depth = 10 m)	T_2 (true depth = 15 m)
S_1 (predicted depth = 10 m)	0.85	0.20
S_2 (predicted depth = 15 m)	0.15	0.80

Should the boring test or the seismic test be carried out by the contractor? What is his optimal strategy?

Solution

The basic decision tree of this problem is shown in Fig. 5.1.

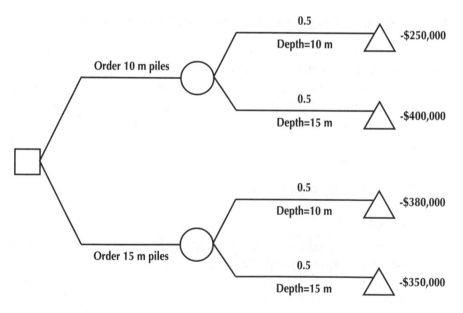

Fig. 5.1 Basic decision tree of the pile ordering problem

The problem is for the contractor to determine if a test should be carried out or not, and if yes, what kind of test is the most cost-effective one.

Consider 3 types of test alternatives:

1. No test (i.e. do-nothing, involves no cost), or
2. Perform a cheaper test (i.e. seismic test, knows sample likelihood only), or
3. Perform a more expensive test (i.e. boring test, knows correct depth).

The seismic test is an imperfect test with a **sample likelihood** as follows:

	T_1 (true depth = 10 m)	T_2 (true depth = 15 m)
S_1 (predicted depth = 10 m)	0.85	0.20
S_2 (predicted depth = 15 m)	0.15	0.80
	$\Sigma=1.00$	$\Sigma=1.00$

This means that if the true state (T_1) that the actual depth is 10 m, there is only a 85% chance that the seismic test can predict S_1 (i.e. predicted depth = 10 m); and if the true state (T_2) that the actual depth is 15 m, there is only a 80% chance that the test can predict S_2 (i.e. predicted depth = 15 m).

If the contractor performs a seismic test, he will face a decision problem as shown in Fig. 5.2. Readers may notice that Fig. 5.2 is derived from the basic decision tree of Fig. 5.1.

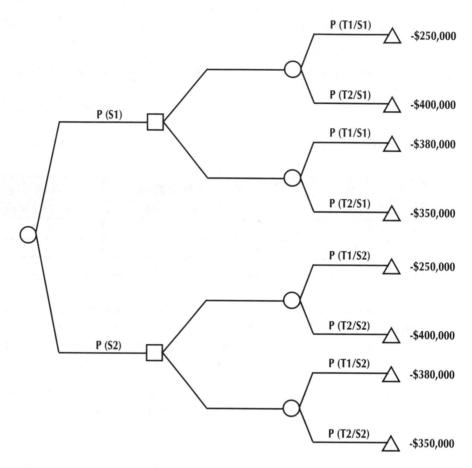

Fig.5.2 Decision tree with conditional probabilities if seismic test is performed

The conditional probabilities shown in the decision tree in Fig. 5.2 are defined as follows:

$P(S_i)$ =   probability of occurrence of S_i under the given sample likelihood with the assumed probabilities of 50:50 chances of 10 m depth and 15 m depth, and

$P(T_j/S_i)$ =   probability of occurrence of T_j when predicted depth S_i occurs.

where i = 1 or 2 and j = 1 or 2

The Bayes' Theorem states that:

$$P(T_j/S_i) = \frac{P(S_i/T_j) \times P(T_j)}{P(S_i)}$$

where $P(S_i/T_j)$ is the sample likelihood given in the problem.

The proof of the above expression (Bayes' Theorem) can be found in most books on basic statistics.

Now, the value of $P(S_i)$ in our problem can be computed as follows:

$P(T_j)$	$P(S_i/T_j)$	$P(S_i \wedge T_j)$

$P(T_1) = 0.5$

$P(S_1/T_1) = 0.85$ — 0.5 x 0.85 = 0.425, i.e. $P(S_1 \wedge T_1)$

$P(S_2/T_1) = 0.15$ — 0.5 x 0.15 = 0.075, i.e. $P(S_2 \wedge T_1)$

$P(T_2) = 0.5$

$P(S_1/T_2) = 0.20$ — 0.5 x 0.20 = 0.100, i.e. $P(S_1 \wedge T_2)$

$P(S_2/T_2) = 0.80$ — 0.5 x 0.80 = 0.400, i.e. $P(S_2 \wedge T_2)$

$P(S_i) \qquad = P(S_i \wedge T_1) + P(S_i \wedge T_2)$

Hence,

$P(S_1) \qquad = P(S_1 \wedge T_1) + P(S_1 \wedge T_2) = 0.425 + 0.100 = 0.525$

$P(S_2) \qquad = P(S_2 \wedge T_1) + P(S_2 \wedge T_2) = 0.075 + 0.400 = 0.475$

The values of $P(T_j/S_i)$ in our problem can also be computed as follows:

$$P(T_1/S_1) = \frac{P(S_1/T_1) \times P(T_1)}{P(S_1)} = \frac{0.85 \times 0.5}{0.525} = 0.81$$

$$P(T_2/S_1) = \frac{P(S_1/T_2) \times P(T_2)}{P(S_1)} = \frac{0.20 \times 0.5}{0.525} = 0.19$$

$$P(T_1/S_2) = \frac{P(S_2/T_1) \times P(T_1)}{P(S_2)} = \frac{0.15 \times 0.5}{0.475} = 0.158$$

$$P(T_2/S_2) = \frac{P(S_2/T_2) \times P(T_2)}{P(S_2)} = \frac{0.80 \times 0.5}{0.475} = 0.842$$

5.2   The Overall Decision Tree for Example 5.1

Now that all conditional probabilities are calculated and known, we are able to draw the **overall decision problem** that the contractor is actually facing. It is shown in Fig. 5.3.

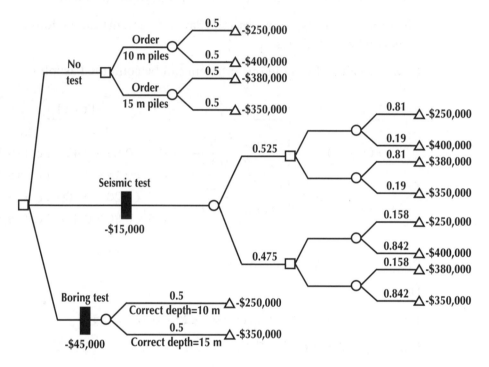

Fig.5.3 The overall decision problem faced by the contractor

In the above decision tree, the EMV of doing nothing (no test) is -$325,000. The EMV after a seismic test is performed is -$329,714 (before deducting the cost of the seismic test is -$314,714). The EMV after a boring test is performed is -$345,000 (before deducting the cost of the boring test is -$300,000)

Since -$325,000 > -$329,714 > -$345,000, therefore doing nothing is the cheapest, then the seismic test, and the most expensive alternative is to perform the boring test. In conclusion, no test should be performed and the contractor should order twenty 10 m piles, a decision from Fig. 5.1.

5.3 The Value of Information

To perform the seismic test or the boring test in Example 5.1 is to provide extra information for the contractor to make his decision. The **value of information** so called is to describe the value of the extra information provided by obtaining the result of the seismic test or the boring test. They can be simply calculated as follows:

Value of seismic test information
= $325,000 - $314,714 = $10,286

If this value is greater than the cost of performing the seismic test, then the test should be performed, and vice versa.

Value of boring test information
= $325,000 - $300,000 = $25,000

If this value is greater than the cost of performing the boring test, then the test should be performed, and vice versa.

In our example, the value of information in each case is smaller than the cost of the test, and therefore no test should be performed.

If the seismic test costs only $9,000 instead of $15,000, then it should be performed because $9,000 is small than $10,286. In such a case, the result of the seismic test prediction (i.e. whether S_1 or S_2) will help us decide what to do, because our decision tree is drawn based on $P(S_1)$ and $P(S_2)$ in Figures 5.2 or 5.3. This is why the Bayes' Theorem and so many probability calculations have been involved. If the seismic test costs $9,000 in this problem, we will perform the seismic test. Then, if the seismic test indicates S_1, we will order 10 m piles; if it indicates S_2, we will then order 15 m piles. That will be the optimum decision overall.

Exercise Questions

Question 1

A company is thinking of selling the development rights of a piece of their land because they think that oil may be found under that land. It will cost the company $150,000 to undertake an exploratory drilling to see whether oil is really present on that site. This money will be wasted if no oil is found. The company's geologist estimates that the probability of presence of oil is 0.60. If oil is present, the company will be able to sell the development rights for $1,000,000.

There can be an alternative besides undertaking an exploratory drilling for finding whether there is oil under the land. The company can carry out seismic tests at a cost of $50,000 for getting information on whether oil is likely to be present or not. It is estimated that if oil is present, there is a probability of 0.85 that the seismic test result is favourable. If, however, oil is not present, there is a probability of 0.1 that the test result is favourable.

What should be the company's optimum strategy?

Question 2

A construction site is about to commence an important piece of excavation work with tight schedule. The work requires a truck to transport the earth excavated. Since the truck is rather old, the project manager is considering whether to do a thorough repair for it before using it for transporting earth.

It costs $10,000 to thoroughly repair the truck. However, if he does not repair it and use it in the excavation work and it then breaks down, it will cost $20,000 to cover the cost of repair and lost time. The project manager estimates that there is a 70% probability that the truck will be OK, but will be 100% sure that it will be OK if it undergoes a repair. Alternatively, a dynamometer test, which costs $1,500, can help test the condition of the engine. This test, however, will only indicate whether the truck is in good or bad condition with a 15% probability the test results being invalid.

What should the project manager do?

6
INVENTORY MODELLING I

6.1 The Nature of the Inventory Problem

A construction firm has to keep a constant inventory of stock. If a site runs out of cement, say, then construction may come to a halt. Conversely, if a site carries an excessive amount of cement, a higher cost will be incurred to store the excessive stock.

Let us consider in more detail the example of storing cement on a site. The decision problem the site agent is faced with is how many bags of cement should be placed in the storage shed. He has to decide how frequent the purchase orders should be made and how many bags should be ordered each time. If he orders a large quantity of cement each time, he has a higher **carrying cost** in storing (or carrying) the cement (i.e. a larger storage space is required). If he orders a smaller quantity each time, he has to place orders more frequently, resulting in a higher **ordering cost**, although he will have a lower carrying cost.

The problem is to determine the minimum total inventory costs (i.e. ordering cost plus carrying cost) so that a balance can be reached between the two costs which have opposite effects on the overall cost.

6.2   Initial Inventory Value and Average Inventory Value

The inventory value (i.e. value of goods stored) varies with time. Its relationship with time is illustrated in Figs. 6.1 to 6.3, where the total inventory consumed is assumed to be in a constant rate in each case. The initial inventory is denoted by I_o which reflects the amount of an order.

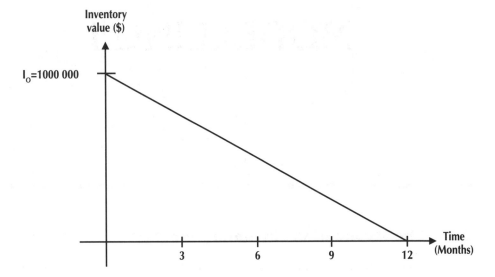

Fig. 6.1 Inventory value graph for N = 1 (N = no of orders in a year). I_o corresponds to the receipt of the order of the goods (the initial inventory value) and the inventory value falls as usage continues. The total inventory value required in a year is assumed to be $1,000,000.

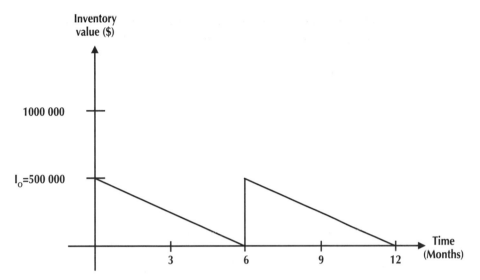

Fig. 6.2 Inventory value graph for N = 2 (i.e. 2 orders in a year). the two peaks of the graph correspond to the receipt of the order. The total inventory value consumed in a year remains to be $1,000,000. I_o (initial inventory) this time is, however, $500,000.

The **initial inventory value** (I_o) is the maximum level of inventory value at the time the order is received.

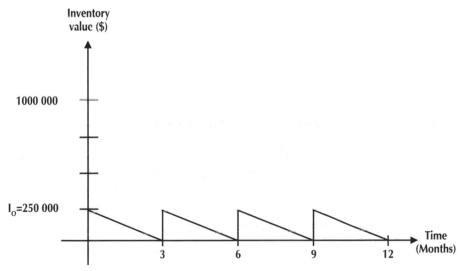

Fig. 6.3 Inventory value graph for N = 4 (i.e. 4 orders in a year). Here the initial inventory value is 250,000 which is attained four times during the year corresponding to the receipt of order.

The **average inventory value** (I_{ave}) can be easily shown to be half of I_o if uniform orders are placed. Consider the example of the case when N = 4.

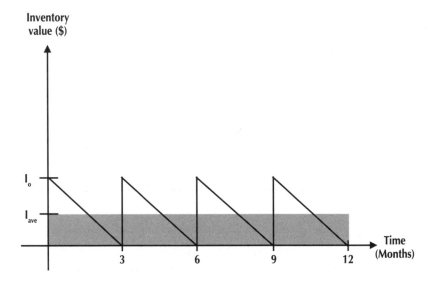

Fig. 6.4 Average inventory value

The average inventory value is defined as the height of the rectangle whose area (shaded in Fig. 6.4) is equal to the total area under the four triangles. It can easily be seen that the average inventory level (I_{ave}) is half of the initial inventory level (I_o). This relationship is true for any value of N.

6.3   Ordering Cost and Carrying Cost

Ordering cost, also known as **procurement cost**, refers to the salaries paid to the staff of the purchasing department, stationery expenses, cost of accounting and administrative work incurred because of purchase orders, office rental and lighting/air-conditioning, transportation of goods, costs incurred in receiving and inspecting the goods, and so on.

The ordering cost is directly proportional to the number of orders made in a given period (usually taken as one year). Graphically, it can be represented as shown in Fig. 6.5.

The slope of the line, P, represents the cost per order ($/order).

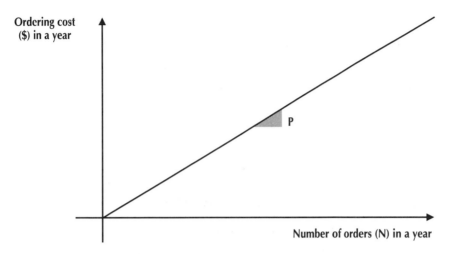

Fig. 6.5 Graph representing ordering cost against N (no. of orders placed in a year) .

Carrying cost, also known as **holding cost**, refers to the rental of the store, costs of lighting and air-conditioning, costs incurred in stock-keeping and security, interest on the lock-up investment capital in inventory purchasing, costs due to goods deterioration, depreciation and insurance, and so on.

Carrying cost is inversely proportional to the number of orders made in a given period. It is represented by a hyperbolic curve, as shown in Fig. 6.6.

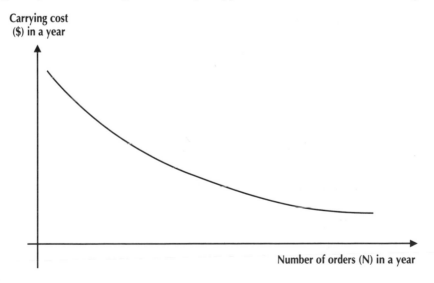

Fig. 6.6 Graph representing carrying cost against N (no. of orders placed in a year)

The hyperbolic characteristic of the curve will be further illustrated in an example in Section 6.5.

For a large inventory value, the carrying cost is high, simply because a large stock has to be kept, and vice verse. In practice, the carrying cost varies directly with the average inventory value (I_{ave}). Therefore, the carrying cost is usually expressed as a percentage of I_{ave}. Carrying cost is usually between 5% to 15% of I_{ave}.

6.4 Economic Order Quantity (EOQ)

The **economic order quantity** (EOQ) is the quantity which should be purchased in an order (I_o) such that the overall inventory cost would be minimized.

The EOQ can be found by a mathematical formula which is derived as follows:

Let
N = number of orders per year
P = ordering cost per order
C = carrying cost per year (expressed as a percentage of I_{ave})
A = total number of items of stock used per year
R = price of each item of stock
T = total inventory cost per year

Then, in any one year,

$$T = \text{ordering cost} + \text{carrying cost}$$

$$= NP + I_{ave}\, C$$

$$= NP + \frac{1}{2} I_o\, C$$

So, $$T = NP + \frac{1}{2}\frac{AR}{N} C$$

To find minimum T, we first have to differentiate T with respect to N.

$$\frac{dT}{dN} = P - \frac{AR}{N^2}\frac{C}{2}$$

T is a minimum when

$$\frac{dT}{dN} = 0 \quad \text{and so,}$$

$$P - \frac{AR}{N^2}\ \frac{C}{2} = 0$$

$$\text{Or,} \quad P = \frac{ARC}{2N^2}$$

$$\text{Hence,} \quad NP = \frac{ARC}{2N}$$

So, ordering cost = carrying cost

Therefore, to get the minimum total inventory cost, the ordering cost should be equal to the carrying cost. The quantity ordered to satisfy this relation is the EOQ.

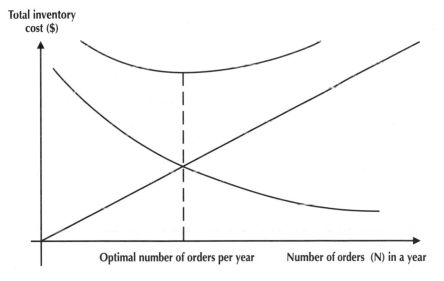

Fig. 6.7   Graph showing total inventory cost. The total inventory cost reaches a minimum value where the two graphs intersect, i.e., when the ordering cost is the same as the carrying cost.

Remember, the quantity of stock ordered for each order in order to get the minimum total inventory cost defines the EOQ. In EOQ, the quantity ordered or number of items per order (i.e. x) is calculated by $x = \sqrt{\frac{2AP}{RC}}$. Also, at EOQ, $N = \sqrt{\frac{ARC}{2P}}$. The proofs are left to the readers. (Hints: A = xN).

6.5 Solving Inventory Problems

We will now illustrate the theoretical derivation given in the previous section by giving an illustrative example.

Example 6.1

A project manager has estimated that he will have to order a total inventory of $2 million in the coming year. From past records he knows that the cost of ordering is $1000 per order and that the carrying cost is 10% of the average inventory value. He wishes to know the optimal number of orders to be placed in the year (i.e. optimal N) so that the total inventory cost is minimized.

He takes several values of N (i.e. 1, 2, 4, 6, 8, 10, 12, 15 and 20) and works out the total inventory cost for each value of N. He uses a table to record his calculations (as shown in Table 6.1)

	(1)	(2)	(3)	(4)	(5)
	Inventory value per order ($) i.e. I_o $= \dfrac{\$2,000,000}{N}$	Average inventory value ($) i.e. I_{ave} $= \dfrac{Column(1)}{2}$	Carrying cost ($) $= I_{ave} \times 0.1$ $= Column(2)$ $\times 0.1$	Ordering cost ($) $= N \times \$1000$	Total inventory cost ($) $= Column(3) +$ $Column(4)$
N = 1	2,000,000	1,000,000	1,000,00	1,000	1,01,000
N = 2	1,000,000	500,000	50,000	2,000	52,000
N = 4	500,000	250,000	25,000	4,000	29,000
N = 6	333,000	166,666	16,666	6,000	22,666
N = 8	250,000	125,000	12,500	8,000	20,500
N = 10	200,000	100,000	10,000	10,000	20,000
N = 12	166,666	83,333	8,333	12,000	20,333
N = 15	133,333	66,666	6,666	15,000	21,666
N = 20	100,000	50,000	5,000	20,000	25,000

Table 6.1 Table showing inventory costs for different N

Table 6.1 is self-explanatory. The minimum total inventory cost occurs when N = 10 (when the carrying cost is equal to the ordering cost).

Notice that the figures in column (1) decrease down the column in a hyperbolic manner, because all the figures in the column when multiplied by N will be equal to the constant value of the total inventory ($2 million). The figures in columns (2) and (3), as in column (1), decrease as N increases in a hyperbolic manner because they are obtained by multiplying the corresponding numbers in column (1) by a constant. Therefore, we can see explicitly that the carrying cost (column (3)) has a hyperbolic relation with N, as mentioned in Section 6.2.

Applying the formulas derived in Section 6.4, we have:

$$N = \sqrt{\frac{ARC}{2P}}$$

$$= \sqrt{\frac{(AR)(C)}{2P}}$$

$$= \sqrt{\frac{2000000 \times 0.1}{2 \times 1000}}$$

$$= \sqrt{100}$$

$$= \underline{\underline{10}}$$

6.6   Inventory Problem: Non-instantaneous Receipt

In all previous problems we have considered, the inventory value of a company varies with time and can be represented by the graph given in Fig 6.8. All along, our assumption is that inventory replenishment is instantaneous.

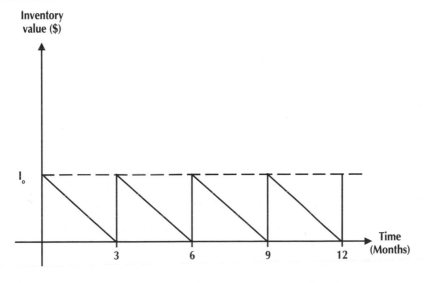

Fig. 6.8   Inventory value graph (instantaneous replenishment)

However, receipts of inventory are not always instantaneous in practice and the process can take days to complete. In such a situation, inventory is being used while new inventory is being delivered. The graph than takes the form as shown in Fig. 6.9.

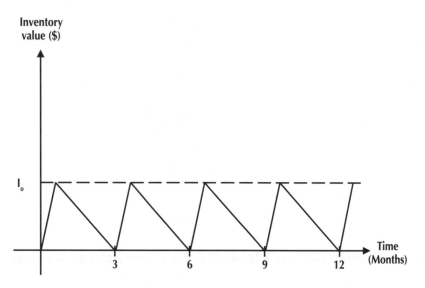

Fig. 6.9 Inventory value graph (non-instantaneous replenishment). New inventory begins when inventory cost reaches zero. The inventory cost then rises because the receipt rate exceeds the usage rate.

Note that even in this situation, when the receipt of inventory is not instantaneous, the average inventory is still equal to half of the maximum inventory. The readers are left to think by themselves why it is so.

We should now determine the optimal size of each order in such a situation.

Let   x = optimal size of order (number of items per order)
       u = usage rate (number of items per day)
       r = receipt rate (number of items per day)

A, P, R and C are defined in Section 6.4. Using the above notation, we can see that the number of days required to receive the new inventory is $\frac{x}{r}$.

So, $\frac{x}{r}u$ is the number of items used during the period of inventory receipt.

Hence, $I_{ave} = \frac{1}{2}(x - \frac{x}{r}u)R$

Carrying cost in a year

$$= I_{ave} C$$

$$= \frac{1}{2} (x - \frac{x}{r} u) RC$$

Ordering cost in a year

$$= \frac{A}{x} P$$

At EOQ, carrying cost = ordering cost, or

$$\frac{1}{2} (x - \frac{x}{r} u) RC = \frac{A}{x} P$$

This reduces to

$$x^2 = \frac{2AP}{(1 - \frac{u}{r}) RC}$$

$$\therefore \quad x = \sqrt{\frac{2AP}{(1 - \frac{u}{r}) RC}}$$

Exercise Questions

Question 1

Prove that the relation $x = \sqrt{\frac{2AP}{RC}}$ given in Section 6.4 is a particular case of the formula derived in Section 6.6.

Question 2

Why carrying cost = ordering cost at EOQ still holds true in the derivation of the formula in Section 6.6 (i.e. for the non-instantaneous receipt condition)? Prove mathematically.

Question 3

(a) Let Q = batch quantity (number of items)
P = cost of placing an order
A = demand in items per annum
R = cost per item
C = carrying cost per annum expressed in % of item cost

Assuming that receipt of inventory is instantaneous, show that the annual inventory cost (T) is given by:

$$T = PA/Q + CRQ/200$$

(b) A contractor requires 1 million bags of cement in a year. Each bag of cement costs $100. It has been estimated that it costs $500 to place an order and that the carrying cost is 10% of item cost per annum. Determine for the contractor, using the formula derived in (a), what quantity should be ordered each time and how frequently orders should be made.

(c) It is found at a later date that the original estimates of the ordering cost and carrying cost percentage were inaccurate and that the true values should be $1,000 and 5% respectively. Find the loss per annum due to the original inaccurate estimate.

7 INVENTORY MODELLING II

7.1 Inventory Modelling Involving Stockout Cost

We have seen in Chapter 6 that under normal circumstances, when inventory is delivered at the instant when an order is placed, the inventory level can be represented graphically as shown in Fig. 7.1.

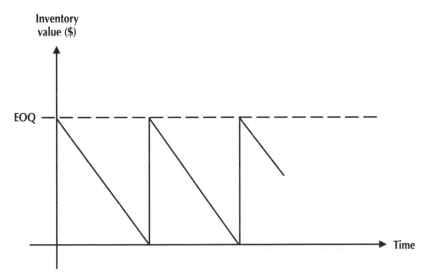

Fig. 7.1   The simple inventory model. Orders are placed at the instant inventory falls to zero. It is then immediately replaced.

However, new stock may arrive later than scheduled, and there may be unexpected excessive demand on stock. Such situations lead to what is known as stockout, which is when the inventory on hand cannot cover needs. This shortage of stock is represented by the shaded portion of the graph under zero inventory in Fig. 7.2.

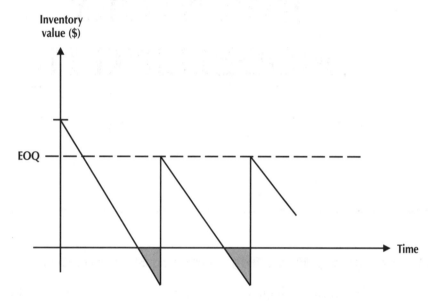

Fig. 7.2   The inventory model with stockout. Stockout is represented by the shaded portion of the graph.

Notice that the new inventory level after stockout does not rise to the original inventory level, because unfilled orders (back orders) of stock have to be filled.

Stockout is expensive because operations have to temporarily cease and the firm may suffer a loss in reputation. Stockout cost is usually expressed as an average cost per item out-of-stock per day, taking all the above costs into account.

We can see that the average inventory, however, will be less than half the maximum inventory when there is stockout in a firm. **This means less carrying cost.**

7.2   Incorporating the Stockout Cost in the Inventory Model

In earlier sections, we saw that the total inventory cost is given by the sum of the ordering cost and the carrying cost only. However, if carrying cost is very high, it may be desirable to **create a stockout situation deliberately**. This can reduce the high carrying cost so that the total inventory cost can be further reduced.

In such a case, the stockout creates a **back order**. That is, the customer does not withdraw the order during the time that the required item is out of stock because the firm will fill the order immediately the inventory item arrives.

For such a model, the total inventory cost is the sum of the stockout cost, the carrying cost and the ordering cost. In this section, we will see how to obtain the minimum total inventory cost for such a problem.

Let us consider the inventory of a firm with stockout and back orders. Suppose that the inventory is represented by the graph in Fig. 7.3.

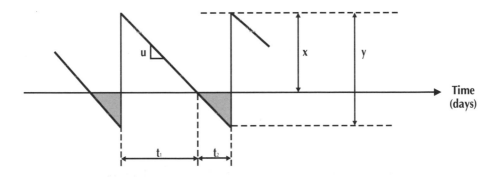

Fig. 7.3   Inventory with stockout and back orders

Let u = usage rate (items per day)
   y = batch order quality
   x = apparent initial stock (after filling back orders)
   t_1 = days when there is stock
   t_2 = days when there is stockout
   s = stockout cost (dollars per item per day)
   h = carrying cost (dollars per item per day)

Stockout Cost

The shaded area in Fig 7.3 represents the number of 'item-days' that items are out of stock. When this is multiplied by the stockout cost per item per day, the total stockout cost is determined.

Total stockout cost for an **inventory cycle** (the period of items between two arrival of new stocks) is given by the shaded area in Fig. 7.2 multiplied by the stockout cost and is given by:

$$= \frac{1}{2}(y - x)t_2 s$$

Now, the total number of days in an inventory cycle is $t_1 + t_2$

Therefore, the average stockout cost per day is given by:

$$\frac{\frac{1}{2}(y - x)t_2 s}{t_1 + t_2}$$

Since $t_1 = \dfrac{xt_2}{y - x}$ (using similar triangles),

$$\text{Stockout cost per day} = \frac{\frac{1}{2}(y - x)t_2 s}{\dfrac{xt_2}{y - x} + t_2}$$

$$= \frac{(y - x)^2 s}{2y}$$

Carrying Cost

The carrying cost can be obtained in a similar manner as follows.

Carrying cost for an inventory cycle $= (\frac{1}{2} xt_1)h$

$$\text{Carrying cost per day} = \frac{\frac{1}{2} xt_1 h}{t_1 + t_2}$$

$$= \frac{\frac{1}{2} x(\dfrac{xt_2}{y - x})h}{\dfrac{xt_2}{y - x} + t_2} \qquad (\text{since } t_1 = \dfrac{xt_2}{y - x})$$

$$= \frac{x^2 h}{2y}$$

Ordering Cost

Let P be the ordering cost per order (i.e. ordering cost in one inventory cycle).

Then, ordering cost per day $= \dfrac{p}{t_1 + t_2}$

$$= \dfrac{P}{\dfrac{y}{u}} \qquad (\text{since } t_1 + t_2 = \dfrac{y}{u})$$

$$= \dfrac{Pu}{y}$$

Minimum Inventory Cost

Let T be the total inventory cost per day.

T = stockout cost per day + carrying cost per day + ordering cost per day

$$= \dfrac{(y - x)^2 s}{2y} + \dfrac{x^2 h}{2y} + \dfrac{Pu}{y}$$

Minimum T is found by solving $\dfrac{\partial T}{\partial x} = 0$ and $\dfrac{\partial T}{\partial y} = 0$.

For $\dfrac{\partial T}{\partial x} = 0$,

$$\dfrac{\partial T}{\partial x} = \dfrac{(2x - 2y)s}{2y} + \dfrac{2xh}{2y} = 0$$

i.e. $x = \dfrac{xy}{h + s}$ .... (1)

For $\dfrac{\partial T}{\partial y} = 0$,

$$\dfrac{\partial T}{\partial y} = \dfrac{2ys(2y - 2x) - 2(y - x)^2 s}{4y^2} - \dfrac{2x^2 h}{4y^2} - \dfrac{Pu}{y^2} = 0$$

i.e. $y^2 s - x^2 h - x^2 s - 2Pu = 0$ .... (2)

Substituting (1) into (2) and simplifying, we obtain:

$$x = \sqrt{\frac{2Pu}{h}} \sqrt{\frac{s}{h + s}}$$

and $y = \sqrt{\frac{2Pu}{h}} \sqrt{\frac{h + s}{s}}$

Example 7.1

A supplier of precast concrete panels undertakes a contract to supply panels to an estate developer. The rate of supply of panels is agreed to be 50 per calendar day. A penalty clause is written into the contract so that the supplier will be penalized at a rate of $10 per panel per day for failure to deliver as agreed.

The supplier wished to cast the panels in batches. He estimates that the cost of setting up a casting bed each time to prepare a new batch is $1000 and that the cost for keeping stock in a rented yard is $2.50 per panel per day. He does not have to pay for the rental if no panel is kept in the yard.

Upon completion of each batch, the panels are taken to the stockyard (after some have been delivered to fill up for the previous period of stockout). Determine for the supplier the quantity of each batch and how frequent he should carry out the casting of a batch.

Solution

In this problem, suppose

h = carrying cost per panel per day = $2.50
s = stockout cost per panel per day = $10
p = setting up cost per batch = $1000
u = number of panels delivered per day = 50

∴ y = optimal batch quantity

$$= \sqrt{\frac{2Pu}{h}} \sqrt{\frac{h + s}{s}}$$

$$= \sqrt{\frac{2 \times 1000 \times 50}{2.5}} \sqrt{\frac{2.5 + 10}{10}} = \underline{\underline{224}} \text{ panels}$$

$$\text{Frequency of batching} = \frac{y}{u}$$

$$= \frac{224}{50}$$

$$= \underline{\underline{4.5}} \text{ days}$$

Therefore, 224 panels should be cast in every batch and a new batch should be completed every 4.5 days.

The number of panels needed to fill the previous unsatisfied requirement each time a batch is completed

$$= y\text{-}x$$

$$= 244 - \sqrt{\frac{2 \times 1000 \times 50}{2.5}} \sqrt{\frac{10}{2.5 + 10}}$$

$$= \underline{\underline{45}} \text{ panels}$$

7.3   Stockout Model With Non-Instantaneous Receipt

Similar to Section 6.6 of Chapter 6, we can remove the instantaneous replenishment assumption for stockout situations. Let us see Fig. 7.4.

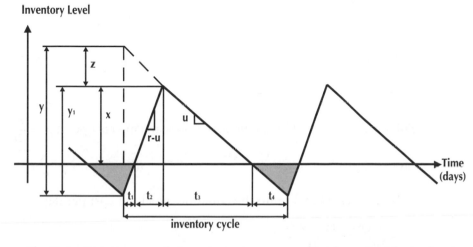

Fig. 7.4   Inventory graph (non-instantaneous replenishment) with stockout

Given u   = usage rate (items per day)

r = receipt rate (items per day)

y = batch order quantity

y_1 = apparent received stock

x = apparent initial stock (after filling back orders and full receipt)

$(t_2 + t_3)$ = days when there is stock

$(t_1 + t_4)$ = days when there is stockout

s = stockout cost (dollars per item per day)

h = carrying cost (dollars per item per day)

P = ordering cost per order

Firstly, we must prove that $y = \dfrac{y_1}{k}$ where $k = (1 - \dfrac{u}{r})$.

From the diagram, the apparent receipt rate = r - u

By the triangles, $t_3 = \dfrac{x}{u}$, $t_2 = \dfrac{x}{(r - u)}$, $t = \dfrac{y}{u}$

$$\text{where } t = t_1 + t_2 + t_3 + t_4$$

Since $z = (t_1 + t_2)\,u$ where $t_1 = \dfrac{(y_1 - x)}{(r - u)}$

So, $z = \left(\dfrac{y_1 - x}{r - u} + \dfrac{x}{r - u} \right) u = \dfrac{u y_1}{(r - u)}$

Hence, $y = y_1 + z$

$$= y_1 + \frac{u y_1}{r - u} = \left(\frac{r}{r - u}\right) y_1 = \frac{y_1}{\left(\dfrac{r - u}{r}\right)} = \frac{y_1}{(1 - \frac{u}{r})} = \frac{y_1}{k}$$

$\therefore y = \dfrac{y_1}{k}$

Carrying cost per day $= \left(\dfrac{hx}{2}\right)\left(\dfrac{t_2 + t_3}{t}\right)$

Substitute $t_2 + t_3$ and t into the equation, carrying cost per day $= \dfrac{hx^2}{2ky} = \dfrac{hx^2}{2y_1}$

Stockout cost per day $= \dfrac{(y_1 - x)s}{2} \dfrac{(t_1 + t_4)}{t}$ where $t_1 = \dfrac{(y_1 - x)}{r - u}$

Substitute t1, t4 and t into the equation, stockout cost per day

$$= \frac{(y_1 - x)^2 s}{2ky} = \frac{(y_1 - x)^2 s}{2y_1}$$

Ordering cost per day $= \dfrac{P}{t} = \dfrac{Pu}{y} = \dfrac{kPu}{y_1}$

Let T = Total inventory cost per day

$$T = \frac{kPu}{y_1} + \frac{hx^2}{2y_1} + \frac{(y_1 - x)^2 s}{2y_1}$$

To find minimum total inventory cost

$$\frac{\partial T}{\partial x} = 0$$

i.e. $x = \dfrac{sy_1}{(h + s)}$ .... (i)

$$\frac{2T}{2y_1} = 0$$

i.e. $(h + s)x^2 = sy_1^2 - 2kPu$ .... (ii)

Substitute (i) into (ii), $y_1 = \sqrt{\dfrac{2kPu}{h}} \sqrt{\dfrac{(h + s)}{s}}$

Since $y = \dfrac{y_1}{k}$, $y = \sqrt{\dfrac{2Pu}{kh}} \sqrt{\dfrac{(h + s)}{s}}$

Substitute y_1 into (i): $x = \sqrt{\dfrac{2kPu}{h}} \sqrt{\dfrac{s}{(h + s)}}$

In summary, we have:

$$x = \sqrt{\frac{2kPu}{h}} \sqrt{\frac{s}{(h + s)}}$$

and

$$y = \sqrt{\frac{2Pu}{kh}} \sqrt{\frac{(h + s)}{s}}$$

Example 7.2

A material supplier undertakes a contract to supply bolts, which are produced by threading standard rods delivered to the shop at a rate of 250 nos. per calendar day, to a building site. The rate of supply of the bolts to the site is agreed to be 100 per calender day. A penalty clause is written into the contract so that the supplier will be penalized at a rate of $10 per bolt per day for failure to supply.

The material supplier wishes to thread the rods in batches. He estimates that the cost of setting up a threading process each time to prepare a whole batch is $2000 and that the cost for keeping stock in the building site where he has rented a portion of land will be equivalent to $5.00 per bolt per day. There will be no charge if no bolt is kept on the site.

Upon completion of each batch, the bolts are taken to the site (after some have been delivered to make up for the previous period of stockout if any). Determine for the supplier the quantity of each batch and how frequent he should carry out the threading process.

Solution

Carrying cost = h = $5 per bolt; stockout cost = s = $10 per bolt

Receipt rate = r = 250 per day; delivery rate = u = 100 per day

$$k = (1 - \frac{u}{r}) = (1 - \frac{100}{250}) = \frac{3}{5}$$

$$y = \sqrt{\frac{2Pu}{kh}} \sqrt{\frac{(h + s)}{s}} = \sqrt{\frac{2 \times 2000 \times 100}{\frac{3}{5} \times 5}} \sqrt{\frac{10 + 5}{10}}$$

$$= \sqrt{200,000}$$

$$= \underline{\underline{447}} \text{ nos}$$

$$t = \frac{y}{u} = \frac{447}{100} = 4.5$$

Frequency of threading process is once in 4.5 days.

7.4   Lead Time and Re-order Level

In practice, there is usually a time lag between the time of re-ordering and the time of arrival of the inventory. This time lag is known as the **lead time**. Therefore, in practice, in order to avoid stockout, the re-ordering must be done before the inventory drops to zero. The inventory level at the time of re-ordering is known as the **re-order level**.

The re-order level is in fact equal to the product of the lead time and the **usage rate** (u). Lead time and the re-order level can be represented graphically, as shown in Fig. 7.5.

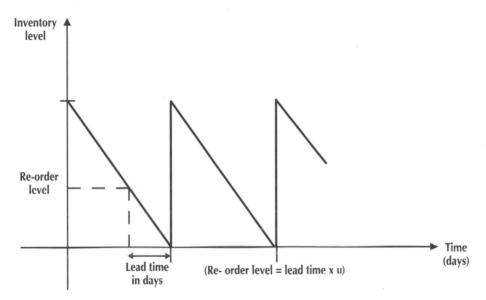

Fig. 7.5 Re-order level and lead time

7.5 Optimal Re-order Level for Uncertain Usage Rate

If the usage rate is constant, it is easy to calculate the re-order level. It is done by multiplying the usage rate by the lead time. However, when the usage rate is not constant (or under uncertainty), the calculation of an optimal re-order level is more complicated. The process is best explained by using an example.

Example 7.3

A contractor has determined, by using the EOQ formula, that at one of his sites, the optimal order quantity of cement is 3650 tonnes per order. The usage of cement on site varies, but the average rate is about 50 tonnes per day. The normal lead time is six days. The stockout cost per tonne of cement per day during the re-order period is $10 and the carrying cost for one tonne of cement per year is $40.

In order to find the optimal re-order level, the firm has investigated the pattern of cement usage from the past records of a similar site. Table 7.1 is a summary of those records.

Tonnes of cement used during the re-order period (lead time)	Number of times this quantity was used
150	0
200	4
250	11
300	72
350	10
400	6
450	3
500	0

Table 7.1 Past record of cement usage during lead time

We can see from Table 7.1 that the most probable usage in the lead time period is 300 tonnes, which is the required amount for the six days lead time.

Determine the optimal re-order level for the contracting firm.

Solution

There are four steps in the solution procedure

Step 1

Establish the probability of various levels of cement usage from Table 7.1. This is shown in Table 7.2.

Cement usage during lead time (tonnes)	Frequency (f)	Probability	Cumulative probability
150	0	0	0
200	4	0.038	0.038
250	11	0.104	0.142
300	72	0.679	0.821
350	10	0.094	0.915
400	6	0.057	0.972
450	3	0.028	1
500	0	0	1

Table 7.2 Probability of cement usage during lead time

From the final column of Table 7.2, we can see that there is:
a 0.962 (i.e. 1 - 0.038) chance of stockout if the re-order level is 200 tonnes;
a 0.858 (i.e. 1 - 0.142) chance of stockout if the re-order level is 250 tonnes;
a 0.179 (i.e. 1-0.821) chance if the re-order level is 300 tonnes, and so on.

Step 2

Calculate the EMVs of stock costs for several assumed re-order levels.

Assumed re-order level (tonnes)	Usage during lead time (tonnes)	Stockout during lead time (tonnes)	Probability (from Step 1)	EMVs of stockout costs (given: stockout cost = $10/tonne/day x 6 days = $60/tonne during lead time)
200	≤ 200	0	0.038	0
	250	50	0.104	50 x 0.104 x $60 = 312
	300	100	0.679	100 x 0.679 x $60 = 4074
	350	150	0.094	150 x 0.094 x $60 = 846
	400	200	0.057	200 x 0.057 x $60 = 684
	450	250	0.028	250 x 0.028 x $60 = 420
	500	300	0	0
			Σ = 1	EMV = $6336
250	≤ 250	0	0.142	0
	300	50	0.679	50 x 0.679 x $60 = 2037
	350	100	0.094	100 x 0.094 x $60 = 564
	400	150	0.057	150 x 0.057 x $60 = 513
	450	200	0.028	200 x 0.028 x $60 = 336
	500	250	0	0
			Σ = 1	EMV = $3450
300	≤ 300	0	0.821	0
	350	50	0.094	50 x 0.094 x $60 = 282
	400	100	0.057	100 x 0.057 x $60 = 342
	450	150	0.028	150 x 0.028 x $60 = 252
	500	200	0	0
			Σ = 1	EMV = $876
350	≤ 350	0	0.915	0
	400	50	0.057	50 x 0.057 x $60 = 171
	450	100	0.028	100 x 0.028 x $60 = 168
	500	150	0	0
			Σ = 1	EMV = $339
400	≤ 400	0	0.972	0
	450	50	0.028	50 x 0.028 x $60 = 84
	500	100	0	0
			Σ = 1	EMV = $84

Table 7.3 EMV of stock out costs
(Note that the sum of probabilities for each re-order level is one.)

Step 3

Calculate the EMVs of carrying costs for the assumed re-order levels.

Assumed re-order level (tonnes)	Usage during lead time (tonnes)	Amount needed to be carried (tonnes)	Probability (from Step 1)	EMVs of carrying costs (given: carrying cost = $40/tonne/year)
200	≥ 200	0	1	0
250	200	50	0.038	50 x 0.038 x $40 = 76
	≥ 250	0	0.962	0
			Σ =1	EMV = $76
300	200	100	0.038	100 x 0.038 x $40 = 152
	250	50	0.104	50 x 0.104 x $40 = 208
	≥ 300	0	0.858	0
			Σ =1	EMV = $360
350	200	150	0.038	150 x 0.038 x $40 = 228
	250	100	0.104	100 x 0.104 x $40 = 416
	300	50	0.679	50 x 0.679 x $40 = 1358
	≥ 350	0	0.179	0
			Σ =1	EMV = $2002
400	200	200	0.038	200 x 0.038 x $40 = 304
	250	150	0.104	150 x 0.104 x $40 = 624
	300	100	0.679	100 x 0.679 x $40 = 2716
	350	50	0.094	50 x 0.094 x $40 = 188
	≥ 400	0	0.085	0
			Σ =1	EMV = $3832
450	200	250	0.038	250 x 0.038 x $40 = 380
	250	200	0.104	200 x 0.104 x $40 = 832
	300	150	0.679	150 x 0.679 x $40 = 4074
	350	100	0.094	100 x 0.094 x $40 = 376
	400	50	0.057	50 x 0.057 x $40 = 114
	≥ 450	0	0.028	0
			Σ =1	EMV = $5776

Table 7.4 EMV of carrying costs

Step 4

Use the results obtained in Steps 2 and 3 to calculate the expected total costs (i.e. expected stockout cost + expected carrying cost). The number of orders in a year is given by:

$$\frac{365}{\frac{A}{u}} = \frac{365}{\frac{3650}{50}} = 5 \text{ orders}$$

Assumed re-order level (tonnes)	EMV of stockout cost (from Step 2) in a year ($)	EMV of carrying cost (from Step 3) in a year($)	Total expected cost per year ($)
200	6336 x 5 = 31680	0	31680
250	3450 x 5 = 17250	76	17326
300	876 x 5 = 4380	360	4740
350	339 x 5 = 1695	2002	3697
400	84 x 5 = 420	3832	4252
450	0	5776	5776

Table 7.5 Table showing the calculations of expected total costs. The minimum total costs ($3697) correspond to a re-order level of 350 tonnes.

Therefore, we can conclude that the optimal re-order level is 350 tonnes of cement. The firm should order 3650 tonnes of cement whenever the stock falls to a level of 350 tonnes. This will provide a **service level** of 91.5% (see Step 1).

Exercise Questions

Question 1

A building project calls for a steady supply of 500 concrete pipes of standard size each week. The price of each pipe is $500. The pipe supplier needs to spend on average $100 for each delivery and the cost of carrying a pipe per week is 5% of the price of the pipe. When the delivery date cannot be met, special arrangements have to be made at the building site. The extra cost incurred in such a situation is $100 for each pipe per week. Calculate the EOQ for the supplier and find out how often batches/deliveries should be made.

Question 2

A local building sub-contractor is engaged in a sub-contract of one year duration to lay the brickwork for some single-storey houses in a low density residential development. Due to some restrictions on the transportation arrangement, the ordered bricks can only be delivered to the main contractor's residential development site on the first day of each month and the order placed must be in one or more complete trucks.

Due to the limited work area of the main contractor's site, all delivered bricks have to be stored in the main contractor's specified warehouse next to the site. Also, only one complete truck of bricks can be taken to site for use at a time and has to be used up prior to taking another complete truck. Basically, storing of the bricks in the main contractor's warehouse is free up to the last day of the month in which these bricks are delivered. Thereafter, any complete truck(s) of bricks left unused will be charged a carrying cost which is shown in the following table:

No. of trucks of bricks left unused	1	2	3	4	5
Carrying cost per truck of bricks per month ($)	800	750	700	650	600

Similarly, the ordering cost of any particular order is dependent on the number of trucks of bricks ordered as follows:

No. of trucks of bricks ordered	5	6	7	8	9	10
Ordering cost ($)	600	700	850	950	1,000	1,000

From past records of similar activity, it has been found that the brick-laying gang of this sub-contractor can place 5 to 10 trucks of bricks in any month with the following probability distribution:

No. of trucks of bricks placed	5	6	7	8	9	10
Probability	0.10	0.15	0.25	0.25	0.20	0.05

It is also known that the average idling cost for the bricklayer is $1,500 per truck of bricks per month.

Find the EOQ (economic order quantity) for the sub-contractor.

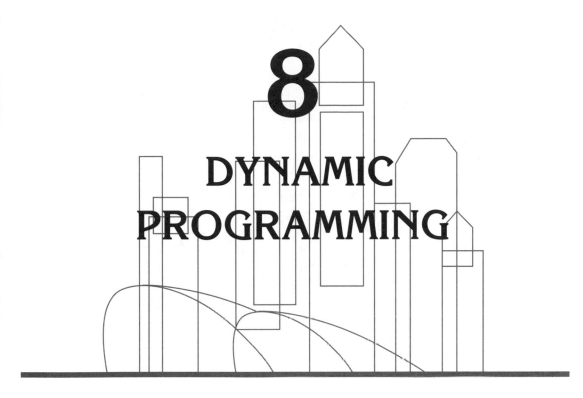

8 DYNAMIC PROGRAMMING

8.1 What Is Meant by "Dynamic"?

Dynamic programming is an optimization technique. The word **dynamic** is used because in this technique decisions are taken at distinct stages. It is based on the principle of optimality, as stated by Richard Bellman (Bellman, 1957) that the overall optimal solution contains the sub-optimal solution that is from the start to a certain stage of a problem.

Let us take a shortest route problem as an example to illustrate what the above statement means, that is, how sub-optimization at an intermediate stage can lead to obtaining the overall optimal solution. Example 8.1 shows the process of finding the shortest route from an origin (Node 1) to a destination (Node 11) using dynamic programming technique.

Example 8.1

Find the shortest route between Node 1 and Node 11 of the following network.

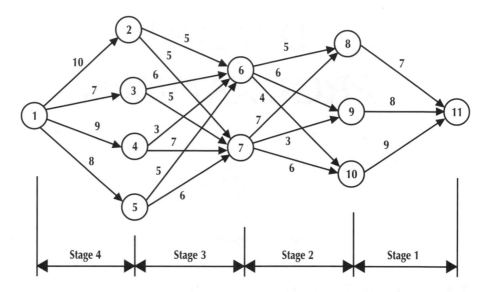

Solution (Method A)

Applying the principle of optimality said above, the shortest route from Node 1 to Node 11 always contains the shortest route from Node 1 to any of the intermediate nodes at a stage on the overall shortest route. In this case, the shortest route from Node 1 to any of the intermediate nodes at that stage is a sub-optimal solution. It is similarly true that the shortest route from Node 1 to Node 11 always contains the shortest route from Node 11 to any of the intermediate nodes at a stage on the overall shortest route, and in this case, the shortest route from Node 11 to any of the intermediate nodes at that stage is a sub-optimal solution. The former case is called the forward method and the latter the backward method.

In solving this problem, we can work either from the beginning to the end (forward method) or from the end to the beginning (backward method). This problem is one of the examples that can be solved by either method. There are dynamic programming problems which **cannot** be solved by the forward method (e.g. Example 8.4 of this chapter). Therefore, in order to play safe, we always use the backward method to solve dynamic programming problems because the backward method is always successful.

In this example, we solve the problem by working backward, starting from node 11.

Starting from node 11, we consider the problem in 4 stages as given below:

Stage 1: Which node should we choose, node 8, or node 9, or node 10 ?

Stage 2: Which node should we choose, node 6, or node 7 ?

Stage 3: Which node should we choose, node 2, or node 3, or node 4, or node 5 ?

Stage 4: Which route is the best route to node 1 ?

Stage 1

If the shortest route passes node 8, then

Distance from node 8 to node 11 = 7

If the shortest route passes node 9, then

Distance from node 9 to node 11 = 8

If the shortest route passes node 10, then

Distance from node 10 to node 11 = 9

Stage 2

If the shortest route passes node 6, then

Distance from node 6 to node 8 then to node 11 = 5 + 7 = 12

Or   Distance from node 6 to node 9 then to node 11 = 6 + 8 = 14

Or   Distance from node 6 to node 10 then to node 11 = 4 + 9 = 13

Hence, the shortest distance from node 6 to node 11 is through node 8 = 12

If the shortest route passes node 7, then

Distance from node 7 to node 8 then to node 11 = 7 + 7 = 14

Or   Distance from node 7 to node 9 then to node 11 = 3 + 8 = 11

Or   Distance from node 7 to node 10 then to node 11 = 6 + 9 = 15

Hence, the shortest distance from node 7 to node 11 is through node 9 = 11

Stage 3

If the shortest route passes node 2, then

Distance from node 2 to node 6 then to node 11 = 5 + 12 = 17

Or Distance from node 2 to node 7 then to node 11 = 5 + 11 = 16

Hence, the shortest distance from node 2 to node 11 is through node 7 = 16

If the shortest route passes node 3, then

Distance from node 3 to node 6 then to node 11 = 6 + 12 = 18

Or Distance from node 3 to node 7 then to node 11 = 5 + 11 = 16

Hence, the shortest distance from node 3 to node 11 is through node 7 = 16

If the shortest route passes node 4, then

Distance from node 4 to node 6 then to node 11 = 3 + 12 = 15

Or Distance from node 4 to node 7 then to node 11 = 7 + 11 = 18

Hence, the shortest distance from node 4 to node 11 is through node 6 = 15

If the shortest route passes node 5, then

Distance from node 5 to node 6 then to node 11 = 5 + 12 = 17

Or Distance from node 5 to node 7 then to node 11 = 6 + 11 = 17

Hence, the shortest distance from node 5 to node 11 is either through node 6 or node 7 = 17

Stage 4

The routes from node 1 to node 11 with distances are:

$$\text{(1)} \longrightarrow \text{(2)} \longrightarrow \text{(11)} \quad = 10 + 16 = 26$$

$$\text{or} \quad \text{(1)} \longrightarrow \text{(3)} \longrightarrow \text{(11)} \quad = 7 + 16 = 23$$

$$\text{or} \quad \text{(1)} \longrightarrow \text{(4)} \longrightarrow \text{(11)} \quad = 9 + 15 = 24$$

$$\text{or} \quad \text{(1)} \longrightarrow \text{(5)} \longrightarrow \text{(11)} \quad = 8 + 17 = 25$$

After performing the above 4 stages of calculations, we can now know the answer by tracing back. The shortest route (with distance) is:

$$\text{Distance} = \text{(1)} \longrightarrow \text{(3)} \longrightarrow \text{(7)} \longrightarrow \text{(9)} \longrightarrow \text{(11)} = 23$$

8.2 Presentation Using Dynamic Programming Language

In the solution given above for Example 8.1, we have not used the standard dynamic programming language to present the calculation process. In the following solution, we will see how the standard form of presentation for solving dynamic programming problems is used. The method is systematic and involves the use of mathematical presentations. The solution for Example 8.1 is re-done for illustration purpose.

Solution (Method B) for Example 8.1

The **return matrix** defined for the problem is:

Destination

		2	3	4	5	6	7	8	9	10	11
	1	10	7	9	8						
	2					5	5				
	3					6	5				
Sources	4					3	7				
(States)	5					5	6				
	6							5	6	4	
	7							7	3	6	
	8										7
	9										8
	10										9

Let $C(S, X_n)$ = the distance of the route when we start at node S and end in an immediate destination node X_n, where S = 1, 2, ..., 10; and n (stage) = 1, 2, 3 and 4.

S is called states and X_n called decision variables. In this problem, the states S for a particular stage are the intermediate nodes at that stage.

$C(S, X_n)$ is shown in the above return matrix.

Define $f_n*(S)$ = the shortest route corresponding to a particular S at stage n

and X_n* = the destination node corresponding to the shortest route at stage n

In the following, we will see how tables are used in the stages to show the step by step calculations using pre-established **return function** and **recursive formula**, which are to be explained in detail.

Stage 1 (n = 1)

S \ X_1	$f_1(S, X_1)$ $X_1 = 11$	$f_1*(S)$	X_1*
S = 8	7	7	11
S = 9	8	8	11
S = 10	9	9	11

Stage 2 (n = 2)

S \ X_2	$f_2(S, X_2) = C(S, X_2) + f_1*(X_2)$ $X_2 = 8$	$X_2 = 9$	$X_2 = 10$	$f_2*(S)$	X_2*
S = 6	5 + 7 = 12	6 + 8 = 14	4 + 9 = 13	12	8
S = 7	7 + 7 = 14	3 + 8 = 11	6 + 9 = 15	11	9

One can see that $f_2(S, X_2) = C(S, X_2) + f1*(X_2)$ is the particular case for Stage 2.

The general case, called the **return function**, is:

$$f_n(S, X_n) = C(S, X_n) + f_{n-1}*(X_n)$$

where $f_n(S, X_n)$ = the total cumulative distance (between the present location S in Stage n and the final destination node 11) and $f_{n-1}*(X_n)$ = the minimum distance from X_n onward to the final destination node 11.

The **recursive function** (or called **recursive formula**) is:

$$f_n*(S) = \min \{ f_n(S, X_n) \}$$
$$= \min \{ C(S, X_n) + f_{n-1}*(X_n) \}$$

Stage 3 (n = 3)

S $\diagdown$ X₃	$f_3(S, X_3) = C(S, X_3) + f_2*(X_3)$		$f_3*(S)$	X_3*
	$X_3 = 6$	$X_3 = 7$		
S = 2	5 + 12 = 17	5 + 11 = 16	16	7
S = 3	6 + 12 = 18	5 + 11 = 16	16	7
S = 4	3 + 12 = 15	7 + 11 = 18	15	6
S = 5	5 + 12 = 17	6 + 11 = 17	17	6 or 7

Stage 4 (n = 4)

S $\diagdown$ X₄	$f_4(S, X_4) = C(S, X_4) + f_3*(X_4)$				$f_4*(S)$	X_4*
	$X_4 = 2$	$X_4 = 3$	$X_4 = 4$	$X_4 = 5$		
S = 1	10 + 16= 26	7 + 16 = 23	9 + 15 = 24	8 + 17 = 25	23	3

Tracing backwards, we can see that the shortest route is 1 - 3 - 7 - 9 - 11 (total distance = 23).

The method of tracing back needs some explanation. At stage 4, S = 1 and $X_4* = 3$, and this means that Node 1 to Node 3 is on the shortest route. Since $X_4* = 3$ at Stage 4, we look for S = 3 at Stage 3. At Stage 3, when S = 3, $X_3* = 7$, and hence Node 7 is on the shortest route. Then we go back to Stage 2, at which when S = 7, $X_2* = 9$. Hence, Node 9 is on the shortest route. At Stage 1, when S = 9, $X_1* = 11$. The shortest route therefore is 1 - 3 - 7 - 9 - 11.

8.3 Further Worked Examples for Dynamic Programming

In this section, we will see how dynamic programming is used to solve problems of different nature.

Example 8.2

A company wishes to take up an opportunity of investing up to $400 million in three projects which are about to commence. They are a housing estate (Project 1), a high-rise commercial-cum-office development (Project 2) and a supermarket-cum-cinema building (Project 3). In whatever way the company invests its money it is guaranteed to receive back the $400 million together with profits in four year's time as given in the following table. All financial quantities are present values.

Investment level (in $50 million's)	Profit Returns (in $ million's)		
	Housing estate	High-rise commercial-cum-office development	Supermarket-cum-cinema building
0	0	0	0
1	5	10	6
2	10	20	12
3	15	30	17
4	20	38	25
5	26	44	34
6	34	47	38
7	40	50	42
8	50	52	48

(a) Use dynamic programming to determine how the $400 million should be allocated, in units of $50 million, among the three projects in order to maximize the total profit.

(b) The company can take up a further option of participating in a development of holiday apartments. This would involve diverting $100 million from the $400 million available for the other three projects but would guarantee the $100 million back together with a profit of $20 million (present value) after four years.

Show that in case (b) the company would now have two optimal policies to choose from and calculate the new maximum profit.

Solution

There should be 3 stages —— 3 different projects (n = 1, 2 and 3)

There are 9 states —— Investment levels (S = 0, 1, 2, 3, 4, 5, 6, 7 and 8)

Decision variables —— X_n = investment level allocated at stage n.

The return function is:

$$f_n(S, X_n) = P_n(X_n) + f_{n-1}^*(S - X_n)$$

The recursive formula is:

$$f_n^*(S) = \max \{ f_n(S, X_n) \}$$
$$= \max \{ P_n(X_n) + f_{n-1}^* (S - X_n) \}$$

Stage 1 (n = 1) —— Consider Project 3 (supermarket-cum-cinema building) only

X_1 S	$f_1(S, X_1) = P_1(X_1)$									$f_1^*(S)$	X_1^*
	$X_1 = 0$	$X_1 = 1$	$X_1 = 2$	$X_1 = 3$	$X_1 = 4$	$X_1 = 5$	$X_1 = 6$	$X_1 = 7$	$X_1 = 8$		
0	0									0	0
1		6								6	1
2			12							12	2
3				17						17	3
4					25					25	4
5						34				34	5
6							38			38	6
7								42		42	7
8									48	48	8

e.g. $f_1(S = 1, X_1 = 1)$ = investing $50 million to supermarket-cum-cinema building, or

$f_1(S = 6, X_1 = 6)$ = investing $300 million to supermarket-cum-cinema building

Stage 2 (n = 2) —— Consider Project 2 (high-rise commercial-cum-office development) <u>AND</u> Project 3 (supermarket-cum-cinema building)

S	$f_2(S, X_2) = P_2(X_2) + f_1{}^*(S - X_2)$									$f_2{}^*(S)$	$X_2{}^*$
	$X_2 = 0$	$X_2 = 1$	$X_2 = 2$	$X_2 = 3$	$X_2 = 4$	$X_2 = 5$	$X_2 = 6$	$X_2 = 7$	$X_2 = 8$		
0	0									0	0
1	6	10								10	1
2	12	10+6 =16	20							20	2
3	17	10+12 =22	20+6 =26	30						30	3
4	25	10+17 =27	20+12 =32	30+6 =36	38					38	4
5	34	10+25 =35	20+17 =37	30+12 =42	38+6 =44	44				44	4 or 5
6	38	10+34 =44	20+25 =45	30+17 =47	38+12 =50	44+6 =50	47			50	4 or 5
7	42	10+38 =48	20+34 =54	30+25 =55	38+17 =55	44+12 =56	47+6 =53	50		56	6
8	48	10+42 =52	20+38 =58	30+34 =64	38+25 =63	44+17 =61	47+12 =59	50+6 =56	52	64	3

e.g. $f_2(S = 3, X_2 = 2)$ = investing $150 million to both projects, of which $100 million is for Project 2, or

$f_2(S = 7, X_2 = 4)$ = investing $350 million to both projects, of which $200 million is for Project 2

Stage 3 (n = 3) —— Consider Project 1 <u>AND</u> (Project 2 <u>AND</u> Project 3)

S	$f_3(S, X_3) = P_3(X_3) + f_2{}^*(S - X_3)$									$f_3{}^*(S)$	$X_3{}^*$
	$X_3 = 0$	$X_3 = 1$	$X_3 = 2$	$X_3 = 3$	$X_3 = 4$	$X_3 = 5$	$X_3 = 6$	$X_3 = 7$	$X_3 = 8$		
S = 8	0+64 =64	5+56 =61	10+50 =60	15+44 =59	20+38 =58	26+30 =56	34+20 =54	40+10 =50	50+0 =50	64	0

e.g. $f_3(S = 8, X_3 = 3)$ = investing $400 million to 3 projects, of which $150 million is for Project 1 (housing estate)

The optimum solution for part (a) of the question is:

Investing no money to housing estate, $150 million (level 3) to high-rise commercial-cum-office development and $250 million (level 5) to supermarket-cum-cinema building. The maximum profit is $64 million.

The above solution is obtained by tracing back the calculation process. At Stage 3, when $S = 8$, $X_3^* = 0$. This means that no money (level $= 0$) should be invested in Project 1. The column for $X_3 = 0$ in the table for Stage 3 is $0 + 64 = 64$, in which 0 represents the profit from Project 1 and 64 (i.e. $64 million) represents the profit from Project 2. If we trace back Stage 2, that $f_2^*(S = 8) = 64$ is corresponding to $X_2^* = 3$. This means that $150 million (level $= 3$) should be allocated to Project 2. The column for $X_2 = 3$ in the table for Stage 2 when $S = 8$ is $30 + 34 = 64$, in which 30 (i.e. $30 million) represents the profit from Project 2 and 34 (i.e. $34 million) represents the profit from Project 3. If we trace back Stage 1, that $f_1^*(S = 5) = 34$ is corresponding to $X_1^* = 5$, which means that $250 million (level $= 5$) should be allocated to Project 3.

In fact, Stage 3 calculation shown above is a short-cut. The original (full) method of calculation for Stage 3 should be as follows:

S	$f_3(S, X_2) = P_3(X_3) + f_2^*(S - X_3)$									$f_3^*(S)$	X_3^*
	$X_3 = 0$	$X_3 = 1$	$X_3 = 2$	$X_3 = 3$	$X_3 = 4$	$X_3 = 5$	$X_3 = 6$	$X_3 = 7$	$X_3 = 8$		
S = 0	0									0	0
S = 1	10	5								10	0
S = 2	20	5+10 =15	10							20	0
S = 3	30	5+20 =25	10+10 =20	15						30	0
S = 4	38	5+30 =35	10+20 =30	15+10 =25	20					38	0
S = 5	44	5+38 =43	10+30 =40	15+20 =35	20+10 =30	26				44	0
S = 6	50	5+44 =49	10+38 =48	15+30 =45	20+20 =40	26+10 =36	34			50	0
S = 7	56	5+50 =55	10+44 =54	15+38 =43	20+30 =50	26+20 =46	34+10 =44	40		56	0
S = 8	64	5+56 =61	10+50 =60	15+44 =59	20+38 =58	26+30 =56	34+20 =54	40+10 =50	50	64	0

e.g. $f_3(S = 8, X_3 = 3)$ = investing $400 million to 3 projects, of which $150 million is for Project 1 (housing estate), or

$f_3(S = 6, X_3 = 4)$ = investing $300 million to 3 projects, of which $200 million is for Project 1 (housing estate)

In part (b) of the question, since $100 million is diverted from the $400 million available for the three projects, we look at the row with S = 6 because this row stands for investing $300 million to the three projects.

For S = 6, $f_3{}^*(S) = 50$ and $X_3{}^* = 0$. This means that besides investing $100 million in the holiday apartments, the other $300 million will be invested in projects 1, 2 and 3, of which zero investment is allocated to project 1 (i.e., housing estate), $200 million or $250 million is allocated to project 2 (i.e. high-rise commercial-cum-office development), and $100 million or $50 million respectively is allocated to project 3 (i.e. supermarket-cum-cinema building). The tracing back method is similar to the one described above.

So, the answer for part (b) is that there are two options for the investment:

Option 1:

$100 million to holiday apartments

$200 million to high-rise commercial-cum-office development

$100 million to supermarket-cum-cinema building

Option 2:

$100 million to holiday apartments

$250 million to high-rise commercial-cum-office development

$50 million to supermarket-cum-cinema building

In both options, the total profit = $50 million + $20 million = $70 million.

Example 8.3

On a construction site, there are 4 stations that will require concrete placing simultaneously tomorrow. 7 foremen will be available tomorrow to supervise these 4 concreting stations. At least 1 foreman must be supervising at each of the stations. Thus, the minimum number of foremen at any one station is 1 and the maximum is 4. The probability that the concreting work will have a bad quality will depend on the number of foremen assigned to the stations as follows:

No. of foremen

	1	2	3	4
Stations 1	0.2	0.15	0.09	0.065
2	0.3	0.2	0.15	0.1
3	0.25	0.18	0.12	0.08
4	0.1	0.08	0.06	0.04

Assuming that the product of the probabilities at the stations will represent the overall probability of having bad quality concrete, use dynamic programming method to find an optimum strategy for assigning foremen to stations tomorrow so that the overall probability of bad quality concreting is minimized.

Solution

Stages —— 4 stations ($n = 1, 2, 3, 4$)

States S —— number of foremen available for assignment at any stage.

Decision variables X_n —— number of foremen assigned to Station $(4 + 1 - n)$

Return function: $f_n(S, X_n) = P_n (X_n) \times f_{n-1}*(S - X_n)$

Recursive formula.

$$f_n*(S) = \min \{ f_n (S, X_n) \}$$

$$= \min \{ P_n (X_n) \times f_{n-1}*(S - X_n) \}$$

Stage 1 ($n = 1$) —— Consider station 4 only

S \ X_1	$f_1 (S, X_1) = P_1(X_1)$				$f_1*(S)$	X_1*
	$X_1 = 1$	$X_1 = 2$	$X_1 = 3$	$X_1 = 4$		
S = 1	0.1				0.1	1
S = 2	0.1	0.08			0.08	2
S = 3	0.1	0.08	0.06		0.06	3
S = 4	0.1	0.08	0.06	0.04	0.04	4

e.g. $f_1(S = 1, X_1 = 1) = 1$ foreman is available and assign this foreman to station 4, or

$f_1(S = 3, X_1 = 2) = 3$ foremen are available and assign 2 foremen to station 4.

Stage 2 (n = 2) —— Consider stations 3 $\underline{AND}$ 4

S \ X_2	$f_2(S, X_2) = P_2(X_2) \times f_1^*(S - X_2)$				$f_2^*(S)$ X_2^*
	$X_2 = 1$	$X_2 = 2$	$X_2 = 3$	$X_2 = 4$	
S = 2	0.25 x 0.1 = 0.025				0.025 1
S = 3	0.25 x 0.08 = 0.02	0.18 x 0.1 = 0.018			0.018 2
S = 4	0.25 x 0.06 = 0.015	0.18 x 0.08 =0.0144	0.12 x 0.1 =0.012		0.012 3
S = 5	0.25 x 0.04 = 0.01	0.18 x 0.06 = 0.0108	0.12 x 0.08 =0.0096	0.08 x 0.1 = 0.008	0.008 4

e.g. $f_2(S = 3, X_2 = 2)$ = altogether there are 3 foremen at stations 3 & 4 but 2 foremen are at station 3, or

$f_2(S = 4, X_2 = 1)$ = altogether there are 4 foremen at stations 3 & 4 but 1 foreman is at station 3.

Stage 3 (n = 3) —— Consider stations 2 $\underline{AND}$ (3 $\underline{AND}$ 4)

S \ X_3	$f_3(S, X_3) = P_3(X_3) \times f_2^*(S - X_3)$				$f_3^*(S)$ X_3^*
	$X_3 = 1$	$X_3 = 2$	$X_3 = 3$	$X_3 = 4$	
S = 3	0.3 x 0.025 = 0.0075				0.0075 1
S = 4	0.3 x 0.018 = 0.0054	0.2 x 0.025 = 0.005			0.005 2
S = 5	0.3 x 0.012 = 0.0036	0.2 x 0.018 = 0.0036	0.15 x 0.025 = 0.00375		0.0036 1 or 2
S = 6	0.3 x 0.008 = 0.0024	0.2 x 0.012 = 0.0024	0.15 x 0.018 = 0.0027	0.10 x 0.025 = 0.0025	0.0024 1 or 2

e.g. $f_3(S = 4, X_3 = 2)$ = altogether there are 4 foremen at stations 2, 3 and 4 but 2 foremen are at station 2, or

$f_3(S = 6, X_3 = 3)$ = altogether there are 6 foremen at stations 2, 3 and 4 but 3 foremen are at station 2

Stage 4 $(n = 4)$ —— Consider stations 1 <u>AND</u> (2 <u>AND</u> (3 <u>AND</u> 4))

S \ X_4	$f_4(S, X_4) = P_4(X_4) \times f_3^*(S - X_4)$				$f_4^*(S)$ X_4^*
	$X_4 = 1$	$X_4 = 2$	$X_4 = 3$	$X_4 = 4$	
$S = 4$	0.2 x 0.0075 = 0.0015				0.0015 1
$S = 5$	0.2 x 0.005 = 0.001	0.15 x 0.0075 = 0.001125			0.001 1
$S = 6$	0.2 x 0.0036 = 0.00072	0.15 x 0.005 = 0.00075	0.09 x 0.0075 = 0.00075		0.00072 1
$S = 7$	0.2 x 0.0024 = 0.00048	0.15 x 0.0036 = 0.00054	0.09 x 0.005 = 0.00045	0.065 x 0.0075 = 0.0004875	0.00045 3

e.g. $f_4(S = 5, X_4 = 1)$ = altogether 5 foremen are at stations 1 & 2 & 3 & 4 but
1 foreman is at station 1, or

$f_4(S = 6, X_4 = 3)$ = altogether 6 foremen are at stations 1 & 2 & 3 & 4 but
3 foremen are at station 1.

The optimum solution is:

Assign 3 foremen in station 1
2 foremen in station 2
1 foreman in station 3
1 foreman in station 4

The above optimum solution is obtained by a tracing back process. At Stage
4, The minimum product of probabilities occurs when $S = 7$ and $X_4 = 3$. This
means that there should be a total of 7 foremen assigned, 3 in station 1. The
column for $X_4 = 3$ in the table for Stage 4 is 0.09 X 0.005, in which 0.005
represents the product of probabilities at Stage 3 for $f_3^*(S = 4)$ and $X_3^* = 2$.
$X_3^* = 2$ means that there should be 2 foremen in station 2. The column for
$X_3 = 2$ when $S = 4$ in the table for Stage 3 is 0.2 X 0.025, in which 0.025
represents the product of probabilities at Stage 2 for $f_2^*(S = 2)$ and $X_2^* = 1$.
$X_2^* = 1$ means that there should be 1 foreman in station 3. Similarly, we can
trace that there should be 1 foreman in Station 4.

Example 8.4

A contract has been signed for the supply of the following number of components at the end of each day to meet a week's demand of a construction site:

Day number	No. of items
1 (Monday)	100
2 (Tuesday)	85
3 (Wednesday)	180
4 (Thursday)	300
5 (Friday)	375
6 (Saturday)	375
7 (Sunday)	285
	1,700 (total)

Production on a day is available for supply at the end of the day, or it may be kept in stock for one night or more at a cost of $1 per item per night. There can be only one batch on a day or no batch at all on the day. If there is a batch on a given day, the batch size can be the supply items of that single day or the supply items of two or more days counting from that given day. If there is no batch on the given day, stocks from the earlier batch/batches of earlier day/days will be used for the supply. The set-up cost (fixed cost) of a batch is $900. The variable cost of production is $2 per item. The problem can be represented by the following diagram:

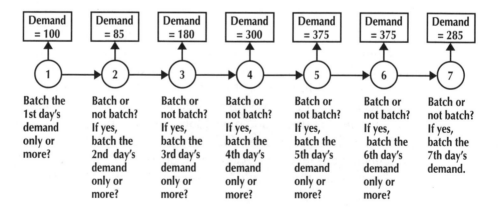

Determine, using the method of Dynamic Programming, on what day/days is/are a batch/batches to be made, and of what size/sizes, if the total cost is to be minimized.

Solution

Stages - 7 days (n = 1, 2, 3, 4, 5, 6 and 7)

States (S) - Day number

Let P_n = Demand (or supply) on the nth day, that is, P_1 = 100, P_2 = 85, ... , P_7 = 285

Decision variables X_n = Number of days in stock between day S and the day of next batch

Recursive formula: $f_n^*(S)$ = minimum $\{f_n(S, X_n)\}$

Return function: $f_n(S, X_n) = 900 + \sum\limits_{k=0}^{X_n} k \times P_{s+k} + f_{n-1-x_n}^*(S - X_n + 1)$

Stage 1: n = 1 (consider day 7)

X_1 / S	$f_1(S, X_1) = 900$ $X_1 = 0$	$f_1^*(7)$	X_1^*
S – 7	900	900	0

e.g. $f_1(S = 7, X_1 = 0)$ = 1 batch (285) on day 7

Stage 2: n = 2 (consider day 6)

X_2 / S	$f_2(S, X_2) = 900 + \sum\limits_{k=0}^{X_2} k \times P_{s+k} + f_{1-x_2}^*(S - X_2 + 1)$		$f_2^*(6)$	X_2^*
	$X_2 = 0$	$X_2 = 1$		
S = 6	900+900 = 1,800	900+1x285 =1,185	1,185	1

e.g. $f_2(S = 6, X_2 = 0)$ = 1 batch (375) on day 6 and 1 batch (285) on day 7, no stock between day 6 and 7

$f_2(S = 6, X_2 = 1)$ = 1 batch (375 + 285 = 660) on day 6 and none later, 285 items of stock between days 6 and 7

Stage 3: n = 3 (consider day 5)

X_3 $f_3(S, X_3) = 900 + \sum\limits_{k=0}^{X_3} k \times P_{s+k} + f_{2-x_3}{}^*(S - X_3 + 1)$ S				$f_3{}^*(5)$ $X_3{}^*$
	$X_3 = 0$	$X_3 = 1$	$X_3 = 2$	
S = 5	900+1,185 = 2,085	900+1x375 +900 = 2,175	900+1x375 +2x285 = 1,845	1,845 2

e.g. $f_3(S = 5, X_3 = 0) = 1$ batch (375) on day 5, some more on day 6 or after

$\quad$ $f_3(S = 5, X_3 = 1) = 1$ batch (375 + 375 = 750) on day 5, some more on day 7

$\quad$ $f_3(S = 5, X_3 = 2) = 1$ batch (375 + 375 + 285 = 1,035) on day 5, and none later

Stage 4: n = 4 (consider day 4)

X_4 $f_4(S, X_4) = 900 + \sum\limits_{k=0}^{X_4} k \times P_{s+k} + f_{3-x_4}{}^*(S - X_4 + 1)$ S					$f_4{}^*(4)$ $X_4{}^*$
	$X_4 = 0$	$X_4 = 1$	$X_4 = 2$	$X_4 = 3$	
S = 4	900+1,845 = 2,745	900+1x375 +1,185 = 2,460	900+1x375 +2x375 +900 = 2,925	900+1x375 +2x375 +3x285 =2,880	2,460 1

e.g. $f_4(S = 4, X_4 = 0) = 1$ batch (300) on day 4, some more on day 5 or after

$\quad$ $f_4(S = 4, X_4 = 1) = 1$ batch (300 + 375 = 675) on day 4, some more on day 6 or after

$\quad$ $f_4(S = 4, X_4 = 2) = 1$ batch (300 + 375 + 375 = 1,050) on day 4, some more on day 7

$\quad$ $f_4(S = 4, X_4 = 3) = 1$ batch (300 + 375 + 375 + 285 = 1,335) on day 4, and none later

Stage 5: n = 5 (consider day 3)

X_5 / S	$f_5(S, X_5) = 900 + \sum_{k=0}^{X_5} k \times P_{s+k} + f_{4-x_5}*(S - X_5 + 1)$					$f_5*(3)$	X_5*
	$X_5 = 0$	$X_5 = 1$	$X_5 = 2$	$X_5 = 3$	$X_5 = 4$		
S = 3	900+2,460 = 3,360	900+1x300 +1,845 = 3,045	900+1x300 +2x375 +1,185 = 3,135	900+1x300 +2x375 +3x375 +900 = 3,975	900+1x300 +2x375 +3x375 +4x285 = 4,215	3,045	1

e.g. $f_5(S = 3, X_5 = 0)$ = 1 batch (180) on day 3, some more on day 4 or after

$f_5(S = 3, X_5 = 1)$ = 1 batch (180 + 300 = 480) on day 3, some more on day 5 or after

$f_5(S = 3, X_5 = 2)$ = 1 batch (180 + 300 + 375 = 855) on day 3, some more on day 6 or after

$f_5(S = 3, X_5 = 3)$ = 1 batch (180+300+ 375+375 = 1,230) on day 3, some more on day 7

$f_5(S = 3, X_5 = 4)$ = 1 batch (180+300+ 375+ 375+285 = 1,515) on day 3, and none later

Stage 6: n = 6 (consider day 2)

X_6 / S	$f_6(S, X_6) = 900 + \sum_{k=0}^{X_6} k \times P_{s+k} + f_{5-x_6}*(S - X_6 + 1)$						$f_6*(2)$	X_6*
	$X_6 = 0$	$X_6 = 1$	$X_6 = 2$	$X_6 = 3$	$X_6 = 4$	$X_6 = 5$		
S = 2	900+3,045 = 3,945	900+1x180 +2,460 = 3,540	900+1x180 +2x300 +1,845 =3,525	900+1x180 +2x300 +3x375 +1,185 = 3,990	900+1x180 +2x300 +3x375 +4x375 +900 = 5,205	900+1x180 +2x300 +3x375 +4x375 +5x285 = 5,730	3,525	2

e.g. $f_6(S = 2, X_6 = 0)$ = 1 batch (85) on day 2, some more on day 3 or after

$f_6(S = 2, X_6 = 1)$ = 1 batch (85 + 180 = 265) on day 2, some more on day 4 or after

$f_6(S = 2, X_6 = 2) = 1$ batch $(85 + 180 + 300 = 565)$ on day 2, some more on day 5 or after

$f_6(S = 1, X_6 = 3) = 1$ batch $(85 + 180 + 300 + 375 = 940)$ on day 2, some more on day 6 or after

$f_6(S = 1, X_6 = 4) = 1$ batch $(85 + 180 + 300 + 375 + 375 = 1,315)$ on day 2, some more on day 7

$f_6(S = 1, X_6 = 5) = 1$ batch $(85 + 180 + 300 + 375 + 375 + 285 = 1,600)$ on day 2, and none later

Stage 7: n = 7 (consider day 1)

X_7 S	$F_7(S, X_7) = 900 + \sum\limits_{k=0}^{X_7} k \times P_{s+k} + f_{5-x_7}*(S - X_7 + 1)$							$f_7*(1)$	X_7*
	$X_7 = 0$	$X_7 = 1$	$X_7 = 2$	$X_7 = 3$	$X_7 = 4$	$X_7 = 5$	$X_7 = 6$		
S = 1	900+3,525 $= 4,425$	900+1x85 +3,045 $= 4,030$	900+1x85 +2x180 $= 4,030$ ... =3,805	900+1x85 +2x180 +3x300 +1,845 $= 4,090$	900+1x85 +2x180 +3x300 +4x375 +1,185 $= 4,930$	900+1x180 +2x180 +3x300 +4x375 +5x375 $= 6,520$	900+1x85 +2x180 +3x300 +4x375 +5x375 +6x285 $= 7,330$	3,805	2

e.g. $f_7(S = 1, X_7 = 0) = 1$ batch (100) on day 1, some more on day 2 or after

$f_7(S = 1, X_7 = 1) = 1$ batch $(100 + 85 = 185)$ on day 1, some more on day 3 or after

$f_7(S = 1, X_7 = 2) = 1$ batch $(100 + 85 + 180 = 365)$ on day 1, some more on day 4 or after

$f_7(S = 1, X_7 = 3) = 1$ batch $(100 + 85 + 180 + 300 = 665)$ on day 1, some more on day 5 or after

$f_7(S = 1, X_7 = 4) = 1$ batch $(100 + 85 + 180 + 300 + 375 = 1,040)$ on day 1, some more on day 6 or after

$f_7(S = 1, X_7 = 5) = 1$ batch $(100 + 85 + 180 + 300 + 375 + 375 = 1,415)$ on day 1, and some more on day 7

$f_7(S = 1, X_7 = 6) = 1$ batch $(100 + 85 + 180 + 300 + 375 + 375 + 285 = 1,700)$ on day 1, and none later

By tracing back, the optimal strategy is:

1 batch of 365 items on day 1, another batch of 675 items on day 4, another batch of 660 items on day 6, and none later.

Total cost = $3,805 + $2 × 1,700 = $7,205

The variable cost of $2 per item is useful only for calculating the total cost. It is not used in performing the dynamic programming calculations.

8.4 Conclusion

We can see from the above examples that, by applying the sub-optimization technique of Bellman's principle of optimality, dynamic programming enables us to greatly reduce the amount of calculations involved for obtaining the overall optimal solution of a problem. Let us look at Example 8.4. In this problem, we have only examine 28 cases (i.e. $1 + 2 + 3 + 4 + 5 + 6 + 7 = 28$). However, if we examine all cases exhaustively, the number of cases would be 64 (i.e. $2^{7-1} = 2^6 = 64$). In general, if this problem involves n days, then by dynamic programming we have to examine $(1 + 2 + ... + n = n(n+1)/2)$ cases, whereas if we examine all cases exhaustively then we have to examine 2^{n-1} cases. So, if the problem involves the supply of items for 30 days (one month), then by dynamic programming we have to examine 465 cases only (i.e. $1 + 2 + ... + 30 = 30 × 31/2 = 465$), whereas if we examine all cases exhaustively we have to examine 536,870,912 cases (i.e. $2^{30-1} = 2^{29} = 536,870,912$). In conclusion, dynamic programming allows us to greatly reduce the number of cases to be examined to look for the optimal solution of a problem. The higher the n is, the greater the reduction will be on the number of cases to be examined. This chapter has shown readers about the theory of it and also the applications.

Acknowledgement

Thanks are due to Mr. H. John Tang who provided valuable contributions to the writing of this chapter.

Exercise Questions

Question 1

A company has six lines on which four different models of PCs (personal computers) can be produced. Due to the differences in variable costs, line efficiency and reject rate, each line provides different profit margins per unit of PCs of each model. For this reason, increasing the number of lines assigned to produce a given model of PC does not proportionally increase the expected total profit. The profit-per-month figures for increasing the number of lines assigned to each model of PC assembly are given in the following table. How many lines, out of six, should be allocated to each model of PC so that the monthly total profit can be maximized?

| No. of lines assigned | Monthly profit | | | |
	PC Model A	PC Model B	PC Model C	PC Model D
0	0	0	0	0
1	40	20	60	20
2	60	40	80	30
3	70	60	80	40
4	70	80	80	40
5	70	90	80	40
6	70	100	80	40
	(Stage 4)	(Stage 3)	(Stage 2)	(Stage 1)

Question 2

A company has an assembly line for one of its products as shown. There are four inspection points as shown in various positions on the assembly line.

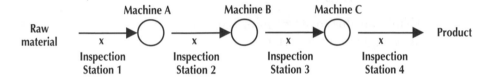

The firm has a total of six quality control inspectors to assign to the four inspection stations; at least one inspector must be stationed at each of the four stations. The probability that a defective product will pass through an inspection station, undetected, depends on the number of inspectors assigned to the station as follows:

Inspection Station	Number of inspectors		
	1	2	3
1	0.3	0.2	0.1
2	0.4	0.3	0.1
3	0.3	0.15	0.05
4	0.2	0.1	0.1

Assume that the probability that a product will pass through all four inspection stations without detection of a defect is equal to the product of the probabilities at each inspection station.

Develop an optimum strategy for assigning inspectors to the inspection stations so that the total probability that a defective product passes through all four inspection stations undetected is minimized.

Question 3

Use dynamic programming technique to solve the following "knapsack problem": A truck can be loaded with a maximum capacity of 51 tonnes. There are seven items of cargo with different weights and values as shown below.

Cargo	1	2	3	4	5
Weight (tonnes)	45	20	30	13	6
Value ($ x 1,000)	20	12	14	3	1

How should the truck carry a combination of the cargo so that the total value of the cargo loaded is maximized and the maximum capacity of the truck is not exceeded for the cases of:

(a) unrestricted number of each cargo in the combination?

(b) maximum number of each cargo = 1 in the combination?

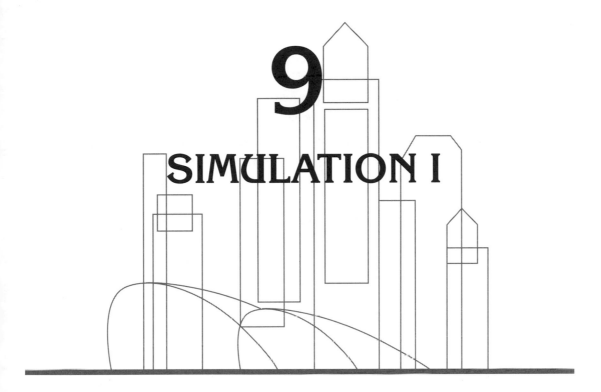

9

SIMULATION I

9.1 Introduction

Simulation is the process of conducting experiments with a model of the system that is being studied or designed. It is a powerful technique for both analyzing and synthesizing engineering and other natural systems.

The simulation procedure is basically an iterative procedure and may be described as an input-output study with feedback provided to guide the changes in the input parameters.

The inputs define the *set of events and conditions* to which the system can be subjected in the real world, and the outputs predict the *system response*. By studying the outputs at the end of each simulation run, one can learn more about the system behavior and may adjust the inputs accordingly.

Simulation models can be broadly grouped into three types:
- Iconic
- Analog
- Analytical

Iconic models are physical replicas of the real systems on a reduced scale. This type of model *is* common in engineering. e.g. In aircraft design, wind tunnels are used to simulate the environment around an aircraft in flight. In the design of large engineering structures, such as skyscrapers, dams, bridges, and airports, *three-dimensional architectural models* are often prepared to provide a realistic view of the design.

A simulation model, in which the real system is modeled through a completely different physical media, is called an **analog model**. In studying the response of engineering structures to various intensities of earthquakes, it is impossible to build a small model of the earthquake zone using rocks and soils and to generate earthquakes at the command of the experimenter. However, if the dynamic property of quake waves is known, an instrument may be constructed to generate a similar type of force motion.

For problems in which the characteristics of the system components and system structure can be *mathematically* defined, then an **analytical model** constitutes a powerful simulation tool. It may be composed of systems of equations, boundary constraints, and heuristic rules, as well as numerical data.

This chapter only focuses on the method of simulation called the **Monte Carlo Simulation**, an analytical model.

9.2   What Is Monte Carlo Simulation?

When we use the word "simulation", we refer to any **analytical method** meant to imitate a real-life system, especially when other analyses are too mathematically complex or too difficult to reproduce.

One type of simulation is **Monte Carlo simulation**, which randomly generates values for uncertain variables over and over to simulate a model.

At the core of simulation is **random number** generation. The computer generates a sequence of numbers, called random numbers or pseudorandom numbers. The numbers generated are considered to be absolutely random and without a pattern. Because of the element of chance, we often call it a Monte Carlo simulation. A Monte Carlo simulation is therefore a probabilistic model involving an element of chance and, hence, it is not deterministic.

The random behavior in games of chance is similar to how Monte Carlo simulation selects variable values at random to simulate a model. When we roll a die, we know that either a 1, 2, 3, 4, 5, or 6 will come up, but we don't know which for any particular roll. It's the same with the variables that have a known range of values but an uncertain value for any particular time or event (e.g. interest rates, staffing needs, stock prices, inventory, phone calls per minute). An analogy to the above example of rolling a die is the generation of random numbers. We will see in the examples in this chapter how simulations are performed by generating random numbers. However, before we go to the examples, let us take a note of the limitations of the Monte Carlo simulation technique.

9.3   Limitations

Despite the many applications and advantages, the following are some limitations of computer simulations:

* The simulation may be expensive in time or money to develop.

* Simulation by itself is **not** an optimization technique. We have to perform a number of simulations and then choose from their results an optimum solution. Because it is impossible to test all alternatives, we can sometimes only provide good solutions but not an optimum solution.

* Because a simulation is probabilistic involving an element of chance, we should be careful of our conclusions.

* The results may be difficult to verify because often we do not have real-world data.

We are going to see what Monte Carlo simulation is by going through worked examples.

9.4   Worked Examples

Example 9.1

A huge construction company possesses 15 crushing plants in its quarry producing aggregates day and night. Long experience tells that there is a 0.20 probability that a crushing plant will break down during any given hour. A number of gangs of fitters (or repairmen) are therefore needed. Depending on the type of failure, the following data are available concerning the required repair time (Table 9.1):

Probability	Time needed for repair
0.08	0.5 hour
0.06	1.0 hour
0.04	1.5 hour
0.02	2.0 hour

Table 9.1 Repair time and its probability

The construction company wants to determine the number of gangs of fitters needed. One gang of fitters, working 8 hours a day, costs the company $2,500 a day and one hour of loss of production for any one crushing plant costs $10,000. There will be 3 shifts for each gang as the plants are running 24 hours a day. Use Monte Carlo Simulation to determine the number of gangs of fitters needed.

Solution

Step 1

The mechanism for determining how long the repair time takes if a crushing plant breaks down is as follows:

Assume there are 100 random numbers (from 00 to 99) and a random number is generated. If this random number is 00, or 01, or 02, ..., or 79, then there is no break down of the crushing plant. This is consistent with the probability shown in Table 9.1 If the random number drawn is 80, or 81, ..., or 87, then the repair time needed by the crushing plant is 0.5 hour (see Table 9.1 again). Similarly, if the random number is 88, or 89, ... , or 93, then the repair time needed is 1.0 hour; if the number is 94, or 95, ..., or 97, then the repair time is 1.5 hours; if the number is 98 or 99, then the repair time is 2 hours. Usually, we do this in a systematic way as shown in Table 9.2:

Crushing plant	Probability	Cumulative Probability	R.N.
No break down	0.80	0.80	00-79
Break down which requires 0.5 hour repair	0.08	0.88	80-87
Break down which requires 1.0 hour repair	0.06	0.94	88-93
Break down which requires 1.5 hour repair	0.04	0.98	94-97
Break down which requires 2.0 hour repair	0.02	1.00	98-99
	1.00		

Table 9.2 Assign R.N. (i.e. random numbers) for repair times

Step 2

Simulate 'n' hours for each crushing plant and tabulate them. The greater the value of 'n', the more accurate the solution will be. Let us take n = 5, say, and tabulate all the random numbers generated. Altogether, 75 (i.e. 15 × 5) random numbers need to be generated. These numbers are totally random and should not exhibit any form of patterns at all. They are shown in Table 9.3.

Crushing Plant	1st hour	2nd hour	3rd hour	4th hour	5th hour
1	47	08	35	72	29
2	**92**	24	53	60	67
3	48	20	11	**89**	71
4	9	5	52	43	00
5	42	45	67	**91**	41
6	38	**87**	54	40	7
7	07	8	**83**	64	**82**
8	58	11	68	74	05
9	13	**92**	**99**	59	22
10	69	16	63	49	39
11	**89**	16	45	42	62
12	24	18	**91**	46	09
13	65	33	02	60	49
14	9	22	21	15	35
15	37	47	18	37	01

Table 9.3 Random numbers generated as repair times

Step 3

As from Step 2 we came to know that there are 10 break-downs of the crushing plants in the 5 hours simulated, because there are 10 numbers which are equal to or greater than 80 (see Step 1). Therefore, in this step, we generate 10 random numbers, which gives the number of minutes after which, in the given hour, the plant stops.

25	27	32	27	11	23	24	10	30	16

Table 9.4 Random numbers generated as the exact minute a crushing plant stops

Fortunately, all the numbers shown in Table 9.4 are small than 60. Since there are only 60 minutes in an hour, any number generated that is 60 or greater is meaningless. In case a random number generated is equal to or greater than 60, we neglect it and generate a new meaningful random number.

From Table 9.4, the first random number is 25. This means that the first break-down of crushing plant occurs at 0 hr 25 min and it takes 1.0 hour to repair, because a random number = 92 corresponds to a repair time of 1.0 hour (see Table 9.2). Similarly, the second break-down occurs at 0 hr 27 min and it takes 1.0 hour to repair (random number = 89 corresponds to a repair time of 1.0 hour too). The third break-down occurs at 1 hr 32 min and it takes 0.5 hour to repair, and so on.

Step 4

Draw a simulation graph as shown in Fig. 9.1 using the data of Steps 2 and 3.

In Fig. 9.1, the numbers shown at the top of each given hour represent the total number of crushing plants broken down at the same time. They are summarized in Table 9.5 in Step 5.

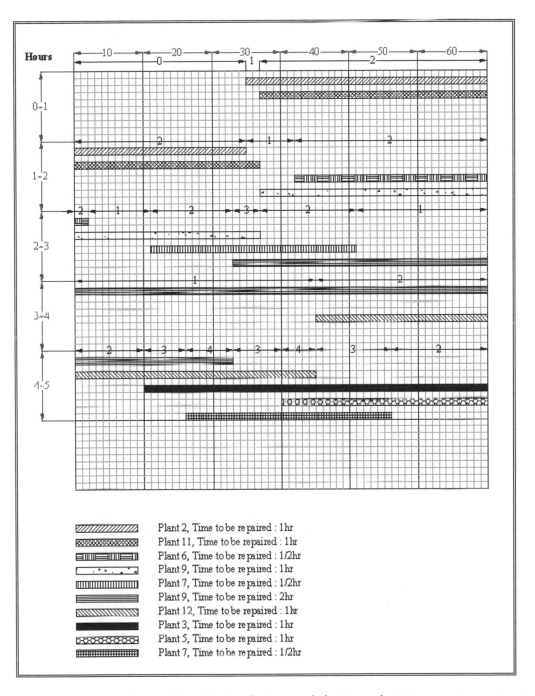

Fig. 9.1 Simulation graph for Example 9.1

Step 5

Summarize the graph in step 4 as shown in Table 9.5.

No, of plants broken down at the same time	No, of minutes	Equivalent hours per day
0	25	24 x (25/290) = 2.07 hrs
1	69	24 x (69/290) = 5.71 hrs
2	161	24 x (161/290) = 13.32 hrs
3	28	24 x (28/290) = 2.32 hrs
4	7	24 x (7/290) = 0.58 hrs
5	0	25 x (0/290) = 0.00 hrs
Total	$\Sigma = 290$	$\Sigma = 24$ hrs

Table 9.5 Summary of breaking down of crushing plants

Step 6

This last step is to determine how many number of gangs of fitters are needed. By inspecting Table 9.5, we need at least 3 gangs (at 3 shifts) or 4 gangs (at 3 shifts). There are 0.58 hr in a day which needs the 4th gang if 3 gangs (at 3 shifts) are employed.

Cost/day to maintain an extra gang (4th gang) at 3 shifts = $2,500 x 3 = $7,500
Cost/day for loss of production if there are only 3 gangs = $10,000 x 0.58 = $5,800

Therefore, it is cheaper to employ 3 gangs of fitters at 3 shifts. It can be concluded that 3 gangs of fitters (at 3 shifts) should be employed.

Note

In Example 9.1, we have performed 5 hours of simulation (n = 5). If we want the result to be more accurate, 100 hours (n = 100) of simulation, say, should be performed. If this is the case, the summary in Table 9.5 will represent a more accurately result.

Example 9.2

Precast concrete piles are taken from the end of a production line into a storage area, from which they are collected by a transporter (having a maximum capacity of 18 piles). The transporter(s) call after a day's production has finished.

The daily production figures were collected, over a period of 100 days, and are given in Figure 9.6 below.

No. of piles produced	9	10	11	12	13	14
No. of days at this production rate	10	18	29	21	12	10

Table 9.6 Production history in the past 100 days

The number of days elapsed, between successive transporter arrivals, is given in Figure 9.7 below.

Days before next transporter arrival	0	1	2	3	4
Frequency of occurrence	32	22	10	20	26

Table 9.7 Interval (days) of transporter arrival

In the above data, days before next transporter arrival = 0 means that two transporters call on the same day.

The storage area can only hold 20 piles. Extra piles have to be stored at a specially arranged place at a cost of $50 for each beam per day (only charged for when used). Alternatively a new storage space (capacity 20 piles also) can be rented for $300 per day. Assume that the storage area is initially empty and that the transporter called on the first day; simulate this system over a period of 10 days. Should the firm rent the new space by using the simulation result?

The following random numbers were generated in sequence for the simulation:

For number of piles produced	12	57	85	78	36	9	60	73	57	86
For days lapsed	29	73	45	58	95					

Solution

Step 1

Tabulate the production figures along with their given probability and calculate the cumulative probability and then assign the random numbers, as shown in Fig. 9.8.

No, of piles produced	No, of days at this production rate	Probability	Cumulative probability	R.N.	No, of piles produced
9	10	0.10	0.10	00-09	9
10	18	0.18	0.28	10-27	10
11	29	0.29	0.57	28-56	11
12	21	0.21	0.78	57-77	12
13	12	0.12	0.90	78-89	13
14	10	0.10	1.00	90-99	14
Total	100	1.00			

Table 9.8 Assign random numbers for the number of piles produced

Step 2

Tabulate the number of days between successive transporter arrivals and assign R.N.

Days before next transporter arrival	Frequency of occurrence	Probability	Cumulative probability	R.N.
0	32	0.29	0.29	00-28
1	22	0.20	0.49	29-48
2	10	0.09	0.58	49-57
3	20	0.18	0.76	58-75
4	26	0.24	1.00	76-99
Total	110	1.00		

Table 9.9 Assign random numbers for intervals (days) of transporter arrival

Step 3

Simulate a period of 10 days:

(i)

DAY	1	2	3	4	5	6	7	8	9	10
R.N.	12	57	85	78	36	9	60	73	57	86
No. of piles produced	10	12	13	13	11	9	12	12	12	13

(ii)

DAY	1	1+1=2	2+3=5	5+1=6	6+3=9
R.N.	29	73	45	58	95
Days Elapsed	1	3	1	3	4

Step 4

Evaluate the simulated result, assuming that the firm has not rented the extra space:

At the end of	Storage (piles)			No. of transporter arrivals
	Before transporter		After	
	arrival	Extra cost		
1st day	10	-	-	1
2nd day	12	-	-	1
3rd day	13	-	-	0
4th day	13+13=26	6x$50=$300	-	0
5th day	26+11=37	-	19	1
6th day	19+9=28	-	8	1
7th day	8+12=20	-	-	0
8th day	20+12=32	12x$50=$600	-	0
9th day	32+12=44	6x$50=$300	26	1
10th day	26+13=39	19x$50=$950		0
		$\Sigma = \$2,150$		

Step 5

Make comparison:

If the firm has rented the extra space, then the amount of rent = $300 \times 10 = $3,000.

Since $2,150 < $3,000, therefore the firm should not rent the extra space.

Example 9.3

Dump truck transport aggregates from an original source to stockpiles. When trucks arrive at the stockpiles they must turn around and then dump the aggregates into the stockpiles from the back. Truck inter-arrival times are distributed according to the following Table 9.10:

Inter-arrival time (min)	4	5	6	7	8	9
Probability	0.10	0.25	0.30	0.20	0.10	0.05

Table 9.10 Truck inter-arrival time and the corresponding probabilities

The times for turn-around and dumping are distributed as shown in Table 9.11:

Turn-around and dumping time (min)	4.0	4.5	5.0	5.5	6.0	7.0
Probability	0.10	0.15	0.35	0.25	0.10	0.05

Table 9.11 Turn-around and dumping time and the corresponding probabilities

Use Monte-Carlo simulation to simulate the arrival and departure of ten trucks starting with the arrival of the first truck. Estimate the mean waiting time of the trucks. The following sequence of random numbers has been generated:

For interval times of trucks	5	65	78	51	23	89	11	63	56	
For turn-around and dumping	69	34	41	10	14	16	8	2	51	7

(Note: Waiting time is the time in the queue and does not include the time taken for turn around and dumping.)

Solution

Step 1

Assign random numbers for inter-arrival times:

Inter-arrival times (min)	Probability	Cumulative Probability	R.N.
4	0.10	0.10	00-09
5	0.25	0.35	10-34
6	0.30	0.65	35-64
7	0.20	0.85	65-84
8	0.10	0.95	85-94
9	0.05	1.00	95-99

The inter-arrival times from the random numbers generated:

R.N.	5	65	78	51	23	89	11	63	56
Inter-arrival times (min)	4	7	7	6	5	8	5	6	6

Step 2

Assign random numbers for turn-around and dumping times:

Turn-around and dumping times (min)	Probability	Cumulative Probability	R.N.
4	0.10	0.10	00-09
4.5	0.15	0.25	10-24
5	0.35	0.60	25-59
5.5	0.25	0.85	60-84
6	0.10	0.95	85-94
7	0.05	1.00	95-99

The turn-around and dumping times from the random numbers:

R.N.	69	34	41	10	14	16	8	2	51	7
Turn-around and dumping times (min)	5.5	5	5	4.5	4.5	4.5	4	4	5	4

Step 3

Draw Fig. 9.2, the simulation result for 10 trucks.

From the simulation, only the second truck has to wait, and it has to wait for 1.5 minutes. The total waiting time is therefore 1.5 minutes. Hence, the average waiting time is 1.5/10 = 0.15.

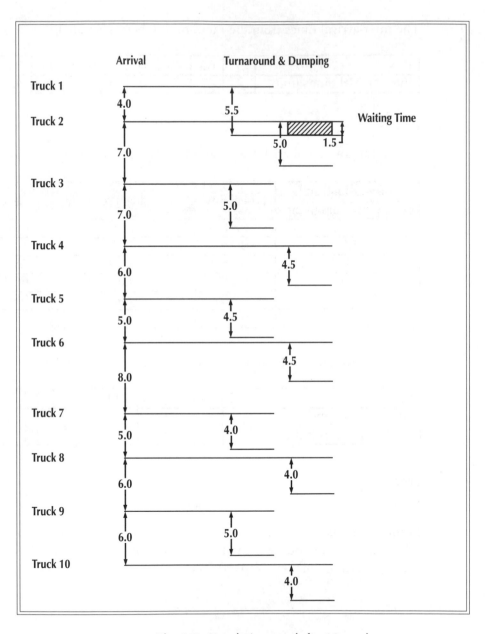

Fig. 9.2 Simulation result for 10 trucks

9.5 Other Applications of Monte Carlo Simulation in Construction

Monte Carlo Simulation is a very versatile method of risk analysis that can be applied to many diverse applications. For the construction industry, Monte Carlo Simulation can be applied to risk analysis of CPM schedules and range estimating.

Monte Carlo Simulation can be achieved in a variety of ways. Applications exist that will perform Monte Carlo within Primavera Project Planner and on their own. Complex Monte Carlo Simulations can also be set up in a spreadsheet application such as Microsoft Excel.

Exercise Questions

Question 1

A ready-mix concrete company supplies building contractors A, B, C and D with units of ready-mix concrete. Contractor A orders 7 units every 4 days; B orders every 5 days but the size of its order varies according to the following table; C orders 5 units whenever it orders, but the interval between successive orders varies according to the following table also:

Size of B's order (units)	P	Interval of C's order (days)	P
5	0.2	3	0.1
7	0.3	4	0.4
9	0.3	5	0.3
11	0.2	6	0.2

Contractor D orders concrete with varying size as well as varying interval, according to the following table:

Size of D's order (units)	P	Interval of D's order (days)	P
3	0.1	2	0.2
6	0.2	3	0.3
9	0.4	4	0.3
12	0.3	5	0.2

Simulate 25 days demand for concrete starting from an initial day when all four contractors order concrete. Obtain, from the simulation, an estimate of the mean demand per day for concrete.

Question 2

The above sketch illustrates an entrance ramp that leads traffic into a single-dual road. The frequencies of arrivals at the ramp-road intersection during peak hours are as follows:

West-bound traffic along highway		Ramp traffic	
Arrival interval (in sec)	Frequency of vehicles	Arrival interval (in sec)	Frequency of vehicles
0	0	0	0
3	50	4	2
6	40	8	3
9	30	12	4
12	30	16	5
15	25	20	6
18	20	24	5
21	15	28	4
24	10	32	3
27	5	36	2
30	1	40	1

Vehicles from the ramp enter into the west-bound lane of the road. To enter safely, a vehicle waiting to enter the road requires 9 seconds before the next vehicle in the west-bound lane arrives. The interval between one vehicle to the next vehicle for entering the road from the ramp is 3 seconds. The problem is to simulate one 10-minute period of traffic flow to determine the average waiting time for the vehicles entering to the road from the ramp. Determine also the maximum number of cars waiting at any one time on the ramp.

Question 3

During the past 10 years there have been extensive flood damages in a city. It is important that one has some means to access flood damage to the city. The following table lists a 10-year record of flood damage:

Year	Discharge, Q (m³ /sec)	Damage, D ($10⁶)
1993	6.5	2.0
1994	9.6	3.4
1995	12.5	5.0
1996	3.7	0.8
1997	5.7	1.5
1998	17.0	6.0
1999	4.8	1.5
2000	9.9	2.8
2001	12.1	5.5
2002	8.5	3.0

Assume that the following relationship applies:

$$D = KQ$$

where D = flood damage in $
Q = discharge rate in m³ /sec
K = multiplication factor, the value of which depends on the time of the flood, concentration & distribution of residential and industrial area, etc.

Simulate a 25-year period and compute the total flood damage for the next 25 years.

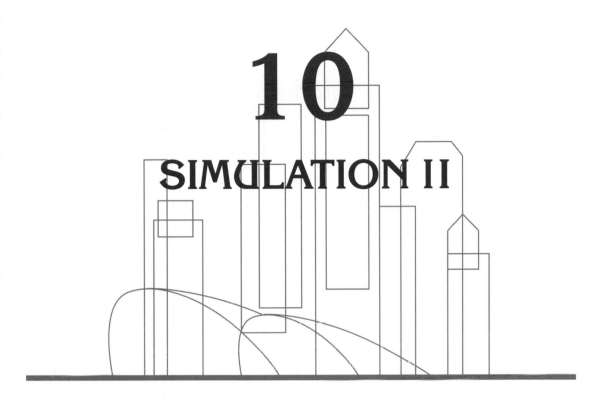

10

SIMULATION II

10.1 Introduction to Simulation Software CYCLONE

Construction planning is the most crucial, knowledge-intensive, ill-structured, and challenging phase in the project development cycle due to the complicated, interactive, and dynamic nature of construction processes (Halpin and Riggs, 1992). The methodology of discrete-event simulation, as discussed in Chapter 9, which concerns "the modelling of a system as it evolves over time by a representation in which the state variables change only at a countable number of points in time" (Law and Kelton, 1982), provides a promising alternative solution to construction planning by predicting the state of a real construction system following the creation of a simulation model based on real life statistics and operations. An event in the context of discrete-event simulation can be defined as an instant of time at which a significant state change occurs in the system (Pidd, 1989). Ever since the inception of the CYCLONE technology (Halpin, 1977), simulation models for typical construction systems have been delivered as electronic realistic prototypes for engineers to experiment on, which eventually will lead to productive, efficient and economical construction operations.

Element Name	Symbol Description	Remarks
NORMAL Activity	Rectangular shape	Unconstrained in its starting logic and indicating active processing of resources entities.
COMBI Activity	Rectangular shape with a slash on its left top corner	Constrained in its starting logic, otherwise similar to the normal modelling element.
QUEUE	"Q" shape; sometimes, a "Q" node is tagged with GEN: N	Representing the waiting or idle state of resource entities; sometimes, the QUEUE node also acts as a GENERATE node to clone resource entities by the number as specified (i.e. N).
COUNTER	Circle with a flag above it	Counting the number of resource entities passing through it.
ARROW	Arrow shape	The resource entity directional flow modelling element
FUNCTION NODE	Circle tagged with GEN:N or CON:N	A number of resource entities as specified (i.e., N) are cloned for a GENERATE Node, or accumulated and merged into one resource entity for a CONSOLIDATE Node

Table 10.1 Basic modelling elements of CYCLONE

CYCLONE uses elements to represent productive/non-productive states of resource entities and portray their dynamic interaction and flow within a construction system. The symbols or modelling elements are designed to be simple and straightforward for developing schematic representations of construction operations. Table 10.1 shows the basic modelling elements of CYCLONE.

10.2 CYCLONE Modelling for Example 9.3

The diagram as shown in Fig. 10.1 is the CYCLONE model for representing and simulating the Truck Dump process of Example 9.3 of Chapter 9. The model was established with the CYCLONE modelling template in Simphony 1.05, which is a construction simulation suite developed by the construction engineering and management program at the University of Alberta, Canada (AbouRizk, 2000).

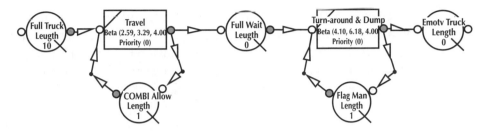

Fig. 10.1 CYCLONE model representing Example 9.3

The arrows in Fig.10.1 show that the activities of the model follow a flow from left to right. The first QUEUE node, "Full Truck", at the leftmost represents 10 trucks carrying aggregates and waiting to travel to the stockpile. Since the problem assumes only one truck can travel to the destination during the process, the "Travel" is described as a COMBI node with another QUEUE node to impose the constraint that the activity "Travel" can only be activated after the previous truck finishes its journey to the dump area. In other words, the resource entity (i.e. truck) undergoes the busy state (as denoted with the "Travel" COMBI node) and the idle state (as denoted with the "COMBI ALLOW" QUEUE node) in processing the ten truckloads of aggregates, which are represented by the "Full Truck" QUEUE node initialized with 10 resource entities. Thus, the QUEUE nodes can be used to represent the *idle* or *waiting* state of resource entities; while the *active* or *busy* state of resource entities are modeled with the COMBI activity nodes. A combination of available resource entities from all the QUEUE nodes linked to a COMBI activity node is the necessary conditions to trigger the start of the activity. For instance, the "COMBI ALLOW" command unit (treated as one resource entity) should be released from processing the previous truckload and become available before getting engaged in processing the next truckload of aggregates waiting in the line.

Once a truck arrives on the dump-site, it needs to wait in front of another activity "Turn-around and Dump". Similarly, since the "Turn-around and Dump" activity should be carried out one truck after another under the direction of a flagman, a COMBI node is required to represent the activity, with a QUEUE node, "FlagMan", acting as a constraint of the limited server capability. Thus, the flagman resource entity should be released from processing the previous truck and become idle before getting engaged in processing the next truck of aggregates waiting in the line. After the completion of dumping, the truck becomes empty and terminates at the last QUEUE node. At the end of simulation, the total number of truck resource entities the system has processed is shown as the length of the "Empty Truck" QUEUE node, which should be initialized with zero length at the start of simulation.

The QUEUE node "Truck Wait" is set initially with length = 0 and this value will increase to the actual queue length (i.e. if the number of truck queuing is 2 then length = 2) during the process of simulation. The QUEUE node "Empty Truck" is set initially with length = 0 but at the end of the simulation, the value will become 10. Unlike Fig. 10.2 (see below), the counter (circle with a flag above it) is not used in Fig. 10.1 because the Simphony template (AbouRizk, 2000) for CYCLONE simulation does not mandate the use of a counter.

10.3 Probability Distribution of Time Duration

The time durations of the two activities "Travel" and "Turn-around and Dump" are transformed into probability distributions. Two beta distributions for "Travel" and "Turn-around and Dump" respectively are input to the COMBI nodes according to the given time and probability. The beta distribution, which was originally used in the classic CPM-based PERT to model activity duration, has been thoroughly studied and conditioned. Not limited to PERT, the generalized beta distribution has also been widely used for modelling the activity duration probabilistically in construction process simulations (AbouRizk *et al.*, 1991) and is given in Equations (1a) and (1b) below.

$$f(x; a, b, L, U) = \frac{\Gamma(a + b)}{\Gamma(a)\,\Gamma(b)} \frac{(x + L)^{a-1}(U - x)^{b-1}}{(U - L)^{a+b-1}} \quad \text{if } L \le x \le U; \quad (1a)$$

$$f(x; a, b, L, U) = 0 \qquad\qquad\qquad\qquad \text{otherwise} \quad (1b)$$

Where Γ = gamma function

$$\text{and} \quad \Gamma(z) \equiv \int_0^\infty t^{z-1} e^{-t} dt \qquad \text{for all } z > 0 \qquad (2)$$

In Equations (1a) and (1b), a and b are the first and second **shape parameters**, and L is the minimum value and U is the maximum value in the probability distribution table. For example, L and U values for the beta distribution of "Travel" duration are 4.00 and 9.00 respectively as given in Table 9.10 of Chapter 9, and those for the beta distribution of "Turn-around and Dump" duration are 4.00 and 7.00 respectively as given in Table 9.11 of Chapter 9. The following paragraphs will discuss shape parameters a and b.

Many statistical and numerical methods have been found to fit a beta distribution and have been documented in literature. Given sufficient sample observations of activity duration, the statistical methods such as moment matching, maximum likelihood, and least-square minimization can be applied to fit beta distributions to the sample data (AbouRizk and Halpin, 1994). When sample observations are not available, subjective information elicited from individuals who are experts on the construction processes can provide reliable clues to statistically represent the activity duration with a beta distribution. Usually, four activity times suffice to define a unique beta distribution numerically. For example, AbouRizk and Halpin (1991) specified the activity's minimum (L) and maximum (U) times together with two of the following: mode, mean, variance, or selected percentiles, to uniquely define a beta distribution. A special case of applying the methods in AbouRizk and Halpin (1991) is to relate the minimum activity duration to the mode, the 75th percentile, and the maximum duration by factors or ratios in fitting beta distributions for truck-traveling times (Fente et al., 2000).

In contrast with the three time estimates in the classic PERT, Lu (2002) proposed a new artificial neural network (ANN) based method of input modelling for PERT simulation requiring four activity-time estimates for each activity, i.e. the minimum (L), the lower quartile (Q_l), the upper quartile (Q_u), and the maximum (U). It is noted that choosing quartiles instead of the mode is mainly due to the fact that it is more straightforward to estimate the quartiles quantitatively from a stream of observed samples. For example, no mode exists by its statistical definition if every observation is unique in the collected data. In case that the sample data are scarce or unavailable, research results in experimental psychology indicates the subjective estimates for the lower, median and upper quartiles can be reasonably accurate (Lichtenstein et al., 1977). Since the ANN engine originally calibrated to enhance the input modelling of PERT simulation is also capable of constructing input models for time/cost variables in construction simulations of any type such as Monte Carlo simulation, a discrete event simulation (Lu 2002), a special program named the *intelligent Simulation Input Modelling Aide* (*iSIMA*) has been developed and used to analyze the concreting productivity data (Lu and Anson, 2004). "*iSIMA*" features (1) the standard statistical analysis utilities and (2) an ANN-based engine for fitting beta distributions which was derived by Lu (2002).

"*iSIMA*" is here used to establish the two beta distributions for the time durations of the two activities "Travel" and "Turn-around and Dump" (i.e. the two COMBI activity nodes of the CYCLONE model shown in Fig. 10.1) based on the time estimates and probability values given Tables 9.10 and 9.11 of Chapter 9. The beta distributions for the "Travel" activity and the "Turn-around and Dump" activity are $BETA(a = 2.59, b = 3.29, L = 4.00, U = 9.00)$, and $BETA(a = 4.10, b = 6.18, L = 4.00, U = 7.00)$ respectively. As explained before, a and b are the two shape parameters of a beta distribution, and L and U are the lower and upper bounds of the given probability distribution.

After inputting the said beta distributions to the model defined in Figure 10.1 and then running the SIMPHONY/CYCLONE simulation model on a computer, the output shows that (1) the maximum queue length for "Turn-around and Dump" is 1, and (2) the queue causes the truck to wait for 1.49 minutes. This is very close to the result obtained in Step 3 of Example 9.3 in Chapter 9.

10.4 A More Complicated CYCLONE Model

A CYCLONE model for concreting operations with crane and skips is given in Fig. 10.2 based on a local building project.

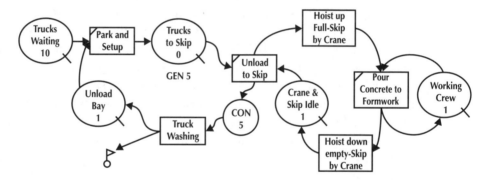

Fig. 10.2 CYCLONE model of Crane and Skip Concreting Operations

Ten truckmixers carrying ready mixed concrete to the site are waiting to be unloaded and processed, as represented with the "Trucks Waiting" QUEUE node at the leftmost. To park a truckmixer and prepare for unloading, an empty unload bay is required, as modeled with the "Unload Bay" QUEUE node initialized with one resource entity representing one unloading bay available on site. The volume capacity of the truckmixers is 5 m³ each, while that of the skip is 1 m³. Hence, each truckmixer is unloaded in five cycles to fill five skip-loads of concrete for pouring on the upper floor location. This is modeled with the "Truck to Skip" QUEUE node with the function of generating resource entities. The tag "GEN 5" means that an arriving resource entity, representing a fully loaded truck, enters into the queue to convert itself into five identical resource entities, representing five skip-loads of concrete. It is important to initialize the QUEUE node with zero resource entities at the start of simulation; the queue length only increments to five once a concrete truck is received on site. To process one skip-load of concrete, the crane together with the skip should be free. So, we must have the "Crane and Skip Idle" QUEUE node and the "Truck to Skip" QUEUE node before the "Unload to Skip" COMBI activity, representing the two resource constraints for triggering the start of "Unload to Skip" activity. The transportation of a skip-load of concrete to the pour location is simply modeled with a normal activity "Hoist up Full-Skip by Crane" that requires no combinational resource constraints. Subsequently, the skipload of concrete will be poured into the formwork by the concreting crew, which is again modeled using a COMBI activity node complemented with a QUEUE node. The crane plus the skip will be released after finishing the pour and returning to the location of the unloading bay (i.e. a resource entity returns to the "Crane and Skip Que" QUEUE node following a simple NORMAL activity of "Hoist down empty skip by crane"). Note that subsequent to the "Unload to Skip" COMBI activity, a function node of consolidating resource entities is inserted in order to count the times a truckmixer is unloaded: once five resource entities are accumulated at the "CON 5" function node, representing a truckmixer being unloaded five times and emptied, then a new resource entity, representing an emptied truckmixer, is released to the "Truck Washing" activity. After that, the truckmixer leaves the site (the counter node following the "Truck washing" activity is used to count the number of truckmixers having been processed by the site system) and the unloading bay is released (one resource entity is available at the "Unload Bay" QUEUE node) for processing the next truckmixer waiting in the queue.

Based on the above CYCLONE description of the concreting operations, the construction manager can readily study the resource configuration and resource utilization of the construction system and estimate the requirement of truckmixer arrival-on-site times so as to achieve the just-in-time concrete delivery and continuous site operations. Of course, the activity duration distributions should be established prior to any simulation experiments. Statistical analysis of simulation outputs based on various sequences of random numbers and multiple simulation runs is needed to reveal and evaluate the system performance. This chapter mainly serves as an introduction to the system simulation modelling and the CYCLONE methodology. Readers may refer to other textbooks or journal papers on construction operations simulation (some are listed in the references below) for more serious treatment of those advanced simulation/statistics topics.

Exercise Questions

Question 1

Use the CYCLONE methodology to develop a schematic model for the construction operations described below:

A road construction contractor adopts the typical road granular base-course construction system featuring typical cyclic and linear processes, which is based on a local road rehabilitation project. Ten rental trucks haul the granular materials from a quarry to the site, the capacity of each being 12 cubic meters. The contractor deploys a shovel at the source area to load dump trucks with granulars and a flagman at the destination area to guide the trucks to dump granulars to specified road sections. The base course to be constructed is 1 km long in total and is divided into 25 sections, each being 40 m long, 12 m wide, and 0.25 m deep. At the site, trucks dump under the guide of a flagman. Once 10 truckloads of granular materials are accumulated for one road section, a grader spreads them to the road profile. Note that after grading one road section, the flagman, who also drives the water truck, will moisten the graded base before a roller compacts it to the designed density. Each road section undergoes the same sequence of activities.

Question 2

The following CYCLONE model describes the construction operations on a local paving job. Explain the construction method and the CYCLONE model in detail. (Hint: the volume capacity of a truck is approximately five times that of the paver's hopper used for paving one section of road surface; the length of one compaction section is about twice that of a paving section.)

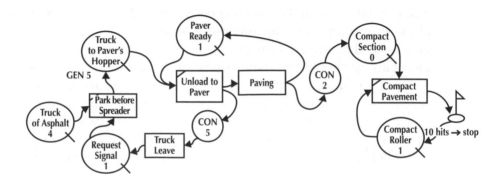

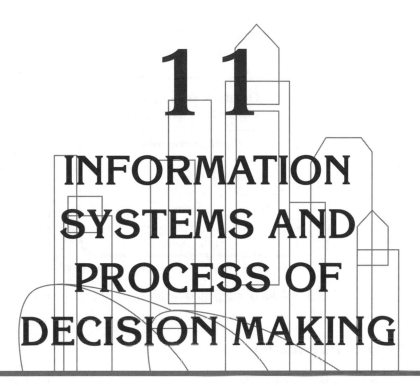

11 INFORMATION SYSTEMS AND PROCESS OF DECISION MAKING

11.1 Introduction

The business of construction is information intensive, dependent on accurate, reliable, up-do-date and timely information. The amount of information can be vast encompassing legal requirements, building codes, specifications and standards, current and historic data about techniques, cost and schedule. Nowadays construction projects are increasingly more complex, and an enormous amount of information need to be processed for effective decision making. Success of a project in today's world is critically dependent on timely and reliable decisions.

Effective decision-making depends on the availability of appropriate information. To facilitate proper and optimal decision making, availability of the desired information at its required level of detail is necessary. Information systems provide the mechanism through which information flows to different departments within and from outside an organization. An information system further facilitates interaction among the managers. Their understanding and control of what is happening in the project improves and they become effective decision-makers (Tenah, 1984). An integrated model of an information system and control environment is illustrated in Fig. 11.1.

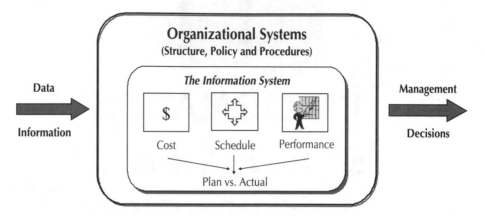

Fig. 11.1 Model of integrated organizational and information system

11.2 **Flow of Information**

Construction is a dynamic process. There are several identifiable stages within this process, such as preliminary design, conceptual design, detail design/ engineering, and construction. The type of work performed during each stage is much different from one another. This is also true about the information that needs to be acquired, generated and processed during the course of each stage. A conceptual flow of information in a typical construction organization is shown in Fig. 11.2.

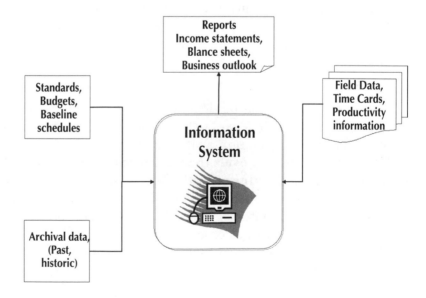

Fig. 11.2 Flow of information in a typical construction organization

It is important to recognize that during preliminary design, not much information is available, although major and critical decisions are made during this stage. Likewise, information during the conceptual design is also extremely critical for success of the project since this information is used to formulate the project basis. At this stage, too, decisions are made on the basis of a limited amount of information. A good information system should provide reliable historic and current information for establishing the budget, the cash flow and the baseline schedule. During detail design and engineering, the system should provide information related to the cost and schedule impact of different design alternatives so that realistic budgets and schedules can be developed based on detailed drawings and specifications.

As soon as construction starts, the amount of available information increases exponentially since resources have been mobilized. The information system should enable the project management team to keep track of all resources (labour, materials and equipment) as they are procured and consumed. One of the major functions of management is to make sure that the resources are available as and when needed. Management must also monitor any deviations from the original plan and take necessary corrective actions if and when necessary based on the available information.

The quantitative techniques presented in Chapters 1 through 10 of this book can be effectively used within an integrated information system.

11.3 Relationship Between Organizational Hierarchy and Information System

The need for information processing depends on the amount of information generated, number of organizational units involved, and interdependency among these units. Construction is a process where a vast amount of information is generated, processed and exchanged; where a large number of inter- and intra-organizational units are involved; and where the interactions among these units are often complex and sometimes confrontational. Therefore, construction organizations and the hierarchy within the organization should be structured in such a way as to satisfy this need. Each level of management in the hierarchy needs appropriate information in a suitable format. Management, depending on its level in the hierarchy, is responsible for making decisions aimed at assuring that the goals and objectives of the project under consideration are met (Kroenke and Hatch, 1994). These decisions may generally be categorized as strategic, tactical, functional, operational and transactional. Executive management makes strategic decisions, whereas middle management makes tactical decisions and functional management makes operational or functional decisions. In Fig. 11.3, a typical organizational hierarchy with corresponding flow of information among different management levels is shown.

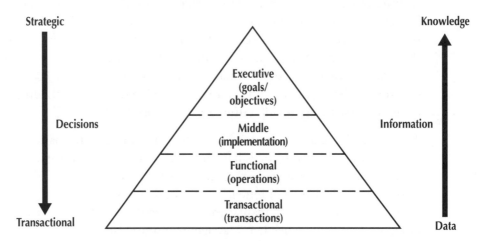

Fig. 11.3   Flow of information in a construction organization

11.3.1 Executive Management Information Needs

The prime decision making focus of executive management is directed mainly towards strategic decision-making. Some of the main informational needs of executive management for their decision-making are listed below:

- Market regulations
- Government regulations
- Historical performance data
- Future prospects
- Financial status/standing of the company
- Budget and schedule at a summary level

Executive management also needs informational feedback on current and forecasted problem areas and recommended procedures for controlling/solving them (Tenah, 1984).

11.3.2 Project/Middle Management Information Needs

The prime decision making focus of project/middle management is directed mainly towards tactical decision-making. Some of the main informational needs of project/middle management for their effective decision-making are listed below:

- Historical performance details (experiential data)
- Resource procurement regulations/conditions
- Resource availability and price trends
- Construction techniques/methodology
- Project participant interaction/inter-relationship
- Operational conditions
- Summary schedule, cost and resource data

Project/middle management also needs informational feedback on estimated versus actual project control parameters (schedule, cost and resource in particular) for them to identify and forecast problems and to recommend counter measures and solution to those problems.

11.3.3 Field/Functional Management Information Needs

The prime decision-making focus of field/functional management is directed mainly towards operational/functional decision-making. Some of the main informational needs of field/functional management for their effective decision-making are:

- Project plans
- Project execution methodology
- Construction techniques and methodology
- Resource procurement, storage and control
- Inspection of schedule, cost and resource status results
- Detailed schedule, cost and resource

Field/functional management needs information to capture detailed work progress data, to identify problematic project area and to recommend solution to these problems.

11.4 Reporting System

The reporting system is very important for project monitoring and control. Not only the executive management but also all project participants need to perceive current project status to catch any errors, which may make cost and schedule deviated from the planned. Project reporting is usually done in the form of progress reports, which should provide adequate information for each management level. A good progress report should have the following characteristics:

- Should be in a standard format and allow readers to quickly grasp the basic information about a project.
- Indicates current project cost, schedule and any variances.
- Forecasts the future trends and provides enough information for decision-making.
- Identify issues and proposes an action plan.
- Should be very brief and requires less than 10 minutes of reading time.

The existing reporting system is in the form of "hard copy" reports, which are prepared and distributed to all the project participants. A common problem with these kind of reports is they contain either "too-little" or "too-much" information. Moreover, it is difficult to update them on a very frequent basis. With the advancement of information technology, it is now possible to produce "electronic reports" that could be made available to all users anytime anywhere. In addition, such reports can filter data according to the need and can be updated almost instantly.

11.5 Data Collection

Data collection is a very important function of an effective project management information system. Not only the decision making but also the successful completion of a project within estimated time and cost depends on the reliability of data collected on-site and off-site.

The different types of data in a typical construction project can be classified as follows:

- **Financial data:** Purchase records; Payment records; Payroll sheets; Bank statements etc.
- **Administrative data:** Employees records; daily, weekly or monthly time sheets; Vendors' and suppliers' information, Contractors' and consultants' information, etc.
- **Field data:** Project status report, Project resources (materials, equipment and labour) report, Time management report (number of hours spent on a particular activity).
- **Technical data:** Contracts and specifications, Project drawings.

Among the above-mentioned data types, the most troublesome and most important is the field data on which all estimation, forecasting and decision-making depends. There are various ways to collect field data like time-, and task-sheets, work and activity sampling, foreman questionnaires, video clips etc. Whatever method is used, the important thing is that the information should be complete, reliable and meaningful. The following guidelines can be helpful for this purpose.

- Keep daily record of all activities performed at the site and the distribution of manpower for these activities.
- Keep daily track of resources (labour, material, equipment) utilization.
- Record everything which can affect project cost and schedule like change orders, bad weather, delayed delivery of material or equipment, labour strike and so on.
- Record any technical or non-technical problems which can affect project performance and report to the main office.
- Always keep backup of important data particularly if the data recording system is computerized.

11.6 Communication

Effective communication is essential for a project to be successful. Communication is the process by which information is exchanged. Information can be communicated by following means:

- Written formal
- Written informal
- Oral formal
- Oral informal

The means used is dependent on the entities communicating. Communication is not simply conveying a message; it is also a process of control and can be used for inciting motivation.

Construction is a multi-organizational process that is heavily dependent on exchange of large amount of complex data and information. Successful completion of a project depends on accuracy, effectiveness and timeliness of communication and exchange of various information and data within the project team.

The availability of modern information and communication tools (ICT) provides the opportunity to project participants to work closely on the project and to improve working environment.

11.7 Use of Information and Communication Technology (ICT) Tools in Construction

Tools and techniques used to gather, store, process, utilize and exchange data and information are generally grouped under information and communication technology (ICT). As in other industries, ICT has brought a revolution in the construction industry. Some representative ICT tools are shown in Fig. 11.4 with their corresponding application areas within the construction project management domain.

Some major recent advances in the field of ICT have made the use and application of ICT tools very convenient. Some of these advances are briefly described below.

11.7.1 Internet

The use of the Internet for information access and financial management purposes independent of time and locational constraints is fast becoming a medium for distance communications and decision making. Now managers can access specific project information in desired formats in an accurate, timely and pertinent form from anywhere across the globe and practically at any time. They can even view live happenings at the site location without physically being at the site, in its city, or even country. Furthermore, any type of information and data can be searched across the globe using World Wide Web (WWW). Programs like Netscape and MS Explorer are available for exploring sites on the web.

11.7.2 Intranet and Extranet

The Intranet is the private web-based network within a corporation. It connects employees and business partners to company information. However, the Intranet applies the WWW model within the boundaries of an organization. Therefore, the Extranet technology is developed to fulfill the Intranet function. An Extranet is a network application that lets companies use the Internet for developing business relationships with partners, suppliers, and customers. Extranets bring competitive advantages by allowing companies to extend internal systems to external business partners. In Fig. 11.5, a schematic view of an online interactive project database system is shown.

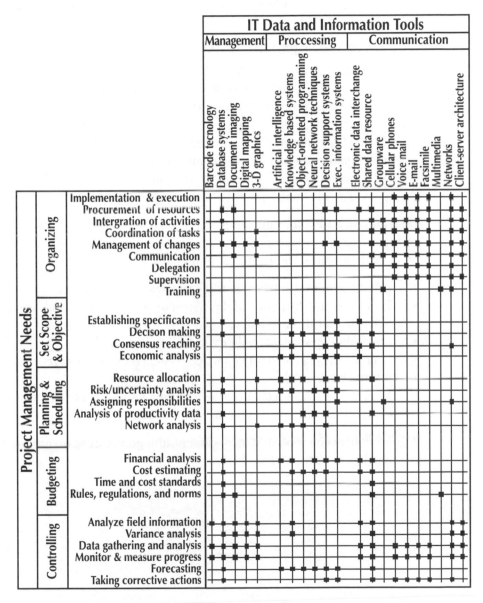

Fig. 11.4 ICT tools and corresponding project management functions

11.7.3 Electronic Mail (E-mail)

E-mail provides a medium to transfer and receive information virtually anywhere in the world. It makes direct communication possible between owners, contractors, suppliers and manufacturers. Technology is available not just to handle text but also images, drawing files, and schedules.

11.7.4 Database Management System

Database management system (DBMS) is a set of software used to develop, implement, manage, and maintain data stored in a database. In other words, it is simply a computerized record keeping system. A DBMS furnishes the means for storing, retrieving, and sorting of data. In construction cost estimating, for example, they provide convenient and systematic access to historical data such as unit prices and productivity rates. In the areas of construction document management, the ability to retrieve and query stored data provides information for claims management.

Information and communication technology is expensive, and before it can be incorporated into any organization, several key factors must be considered from both business and technical points of view such as:

- Cost: *Initial investment versus the savings in terms of reduced man-hours.*
- Flexibility: *the feasibility of the system to accept changes and be adapted to multiple user demand and use.*
- Scalability: *system adaptability to future trends and use.*
- Quality: *the degree to which the ICT system aids the manager in doing his/her work effectively.*
- Importance: *the level of dependence of the project on the ICT system*
- Content: *the accuracy of the data in the system*

A good fit between tools and management functions can be determined using the matrix presented in Fig. 11.4.

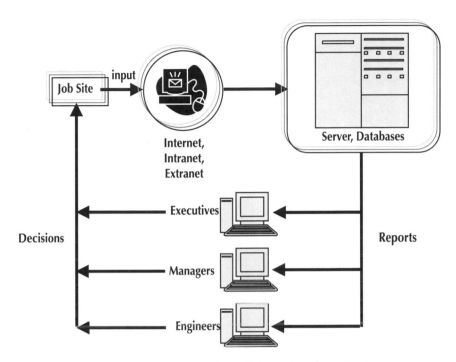

Fig. 11.5 Schematic view of a web-based on-line database system

Exercise Questions

Question 1
Describe the relationship between organizational hierarchy and information systems.

Question 2
"Flow of information is a bottom-up process while decision-making is a top-down process." Explain what kind of problems may occur due to this apparent anomaly in a construction organization and how the problem can be minimized.

Question 3
Briefly explain different kind of data generated and used in the process of construction.

Question 4
Describe the various ways data are communicated. How can be IT tools used to enhance communication of data?

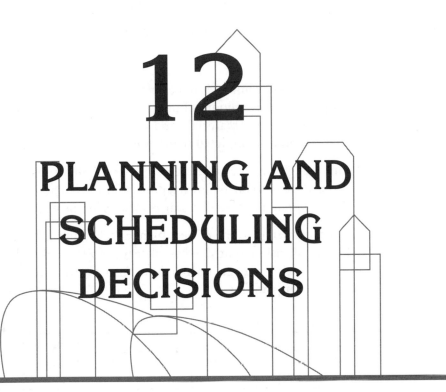

12

PLANNING AND SCHEDULING DECISIONS

12.1 Introduction

Completing a project on time, within budget and according to specifications should be the overall objective of a project team. This objective cannot be achieved without proper planning and scheduling.

Planning and scheduling of the functions, operations and resources of a project are amongst the most challenging tasks faced by a project management team. The goal is to sequence operations properly and to allocate efficiently the resources involved. In this chapter, decision-making aspects of planning and scheduling are discussed. Brief introductions to various planning and scheduling tools are also included. For a comprehensive presentation of these quantitative techniques readers are referred to Tang, *et al.* (2003).

12.2 Planning and Scheduling

12.2.1 Project Planning

Planning involves deciding in advance what is to be done, how and in what order, so that the project can be managed effectively.

- The objective of planning is to be able to complete the project effectively within a reasonable amount of time, using available resources including money, manpower, materials and equipment. Thus a well-planned project would be completed without unusual delay and cost overrun.

- Planning begins with estimating. They are simultaneous processes. Just as it is necessary to break down a project for the purpose of estimating it is also important to consider a project in parts for the purpose of planning.

- The real challenge to the project planner/scheduler is his or her ability to identify all tasks required to complete the project. Being able to identify work activities and grouping them into meaningful categories is important for planning.

Tasks	Challenges	Solutions
• Planning data collection	Where to look for data?	Studying relevant documents
• Planning time	What is to be done?	Defining scope of work
	What are the activities involved?	Breaking down the project into activities
	How can it be done?	Developing sequence, network plans
	When is it to be done?	Scheduling tasks
	Where is it to be done?	Charting site layout
• Planning resources	What is needed to do it?	Forecasting resource requirement
		Planning manpower requirement
		Planning materials procurement
		Planning equipment procurement
		Budgeting costs
	Who is to do it?	Designing organizational structure
		Allocating tasks and resources
		Establishing responsibility centers
• Planning implementation	How to account for performance?	Designing control system
	How to monitor performance?	Formulating monitoring process
	How to communicate information?	Developing project management information system

Table 12.1 Project planning process

- The process of developing a well-defined work breakdown structure (WBS) often leads to a list of activities that must be performed to complete a project.

- Project planning provides the central communication that coordinates the work of all parties. Planning also establishes the benchmark for the project control system to track the quantity, cost, and timing of work required to successfully complete the project.

Example 12.1

Decision making at the estimating stage

Consider the following two line items taken from an estimate worksheet for labour cost only.

Item	Quantity(tonnes)
1. Erect structural steel	280
2. Erect joist steel	200

The estimator/planner must decide one of the following crews, as described below, to use for the two tasks.

Crew 1

1 foreman	@$65.00	$65.00
4 iron-workers	@$50.00	200.00
1 operator	@$45.00	45.00
1 oiler	@$40.00	40.00
2 labourers	@$35.00	70.00
Total 9 crew-hours		$420.00

Unit crew price = $46.67

Crew 2

1 foreman	@$65.00	$65.00
5 iron-workers	@$50.00	250.00
1 operator	@$45.00	45.00
1 oiler	@$40.00	40.00
3 labourers	@$35.00	105.00
Total 11 crew-hours		$505.00

Unit crew price = $45.91

	Production Rates (tonne/day)		Crew hours	
Item	Crew 1	Crew 2	Crew 1	Crew 2
1. Erect structural steel	10.4	12.3	1960*	2003
2. Erect joist steel	14.4	16.1	1000	1093

* Explanation: 10.4/8 = 1.3 tonne/hr; (1/1.3) = 0.79 hr/tonne; (0.79 × 9) = 7 crew-hr/tonne; 7 × 280 = 1960 crew-hours (assuming 8-hour days)

	Cost ($)		$/tonne	
Item	Crew 1	Crew 2	Crew 1	Crew 2
1. Erect structural steel	$91,473**	$91,957	$326.70	$328.50
2. Erect joist steel	$46,670	$50,180	$233.35	$250.90

** Explanation: 1960 × $46.67 = $91,473

	Days needed	
Item	Crew 1	Crew 2
1. Erect structural steel	27***	23
2. Erect joist steel	14	12
Total	41	35

*** Explanation: 1960/(9 × 8) = 27 (rounded)

From the above analysis an estimator/planner can easily decide which crew to employ considering the differences in cost, unit cost, production rate and number of days needed to do the tasks. It is interesting to note that for item 1, *erect structural steel*, Crew 1 and Crew 2 costs are almost the same although there is a total of 4 days saved by using Crew 2.

12.2.2 Project Scheduling

Scheduling aims at formulation of a time-based plan of action for coordinating various activities and resources. A project schedule shows the sequence and interdependencies of activities, their time durations and their earliest and latest completion times.

12.2.3 Key Principles of Planning and Scheduling

- Begin planning before starting work, rather than after starting work
- Involve people who will actually do the work in the planning and scheduling process
- Include all aspects of the project: scope, budget, schedule, and quality
- Build flexibility into the plan, include allowance for changes and time for reviews and approvals
- Remember that the schedule is the plan for doing the work, and it can never be precise

- Keep the plan simple, eliminate irrelevant details that prevent the plan from being readable
- Communicate the plan to all parties; any plan is worthless unless it is known

12.3 Elements of Planning

12.3.1 Summarizing goals and the scope of work

The goal or goals of the project should be clearly understood and agreed on by all planning participants. The basic approach to planning involves segmenting the total endeavor into manageable parts, planning each part in detail, combining the parts and testing the total against project objectives, and then refining the planning as necessary to eliminate variances from the objectives.

Defining the scope of work is of utmost importance, since scope definition provides a means of identifying tasks and activities. The most effective tool to use in ensuring that all work scope is planned is the work breakdown structure (WBS). The WBS is a tree structure of successively further breakdowns of work scope into component parts for planning, assigning responsibility, managing, controlling and reporting project progress. All planning efforts should be organized to the WBS developed for the project.

Planning takes place in numerous categories, but the most important of these are time, cost, resources, and quality.

12.3.2 Time Planning

Time planning entails developing plans to accomplish all elements of an objective within an established time period. These plans are then developed into detail schedules for accomplishing discrete tasks. This process begins with establishing an end date or other milestone dates at which all actions must be complete, and works backward from that point.

Steps

- Divide the total effort into component parts or activities.
- Arrange the activities in the order of their accomplishments considering the relationships among them (some activities must be handled in strict sequence while others may be executed simultaneously, and still for others a number of options may exist.).

- Draw a critical path logic diagram depicting the relationships defined above (in this format, arrows or nodes representing each component or activity are displayed in logical sequence, showing dependencies among all activities.).
- Assign durations (time period) to each activity, based on experience of the planning team and pertinent analysis (see Example 12.1 above).
- Perform critical path calculations in order to establish the project duration, and to define the critical path (chain of critical activities) and slacks in non-critical activities. Commercial software programs are widely used nowadays to carry out the CPM (critical path method) calculations. MS PROJECT and PRIMAVERA PROJECT PLANNER are the two most commonly known packages among many others.
- If the total time exceeds the available time, planners must reevaluate their work and take necessary actions (perhaps shortening some activities by applying more resources).

12.3.3 Cost Planning

Cost planning is based on the cost breakdown structure. Ideally, this segment will parallel the time breakdown structure or the activity breakdown structure. The objective is to establish the time-cost relationships for the planned activities. This, however, may not be entirely possible since not all costs are directly related to a specific activity and also because a logical activity breakdown structure may not be a meaningful cost breakdown structure. But cost and activity planning is the first step in defining work packages that can be used as fundamental planning as well as control units. A work package or a control account is created where costs and actions coincide.

12.3.4 Resource Planning

Resources involved in an undertaking generally include personnel, support equipment and tools, permanent materials and installed equipment, and expendable supplies. Some or all of these resources are needed in each control account as defined above (at the intersection of the activity breakdown structure and cost breakdown structure). The decision as to the resources to be applied is primarily based on experience and judgement, although specific undertakings may require other input as well. Every resource requirement must be accounted for in the cost breakdown so that estimates of costs for individual control accounts, as well as total estimated costs, can be generated. Resource planning includes planning for quantities needed at proper time.

Certain resources are critical to project success, and must be identified with all the necessary information (quantity, time and duration, etc.).

12.3.5 Quality Planning

Quality planning depends on goals set for achieving quality targets. This includes project requirements in the form of specifications, a method for communicating the requirements to those responsible for achieving them, a plan for training the responsible persons, and a way of measuring successful achievement.

12.4 Tools for Planning & Scheduling

There are a number of different analytical tools and graphical techniques for planning and scheduling of construction operations. They include the following:

1. Bar charts
2. Progress curves
3. Linear balance charts
4. Networks

It is important to note that none of the above in and of itself represents a "plan" for the project. The complete plan exists only in the minds of the planners. The tools listed above are merely the means to aid planners in the following ways:

1. Organizing and documenting their thoughts about the plan.
2. Communicating their thoughts to the people who will ultimately put their plan into action.

12.4.1 Bar Charts

A construction project could be considered as a collection of interrelated jobs or activities all aimed at achieving a common goal. A bar chart graphically represents a project's activities or jobs and their timing. The activities in a bar chart are generally listed on the left side of the chart, vertically one below the other (along the y-axis). A horizontal time scale extends to the right of this list (along the x-axis). A horizontal bar is drawn against each activity listed in the chart between the corresponding start and finish times of the activity. Bar charts can also be used to record and report the progress of

Activity No.	Description	Months							
		1.	2.	3.	4.	5.	6.	7.	8.
1.	Footings								
2.	Columns								
3.	Walls								
4.	Slabs								
5.	Doors/Windows								

Fig. 12.1 A bar chart schedule

actual work completed. In order to report progress, a parallel bar is sometimes placed immediately below the plan bar. As the job progresses, it is shaded in direct proportion to the physical work completed. An example of a bar chart is shown in Fig. 12.1.

Advantages of bar charts

One of the major advantages of bar charts over other planning and scheduling tools is their relative simplicity. It is quite easy to develop and understand a bar chart. This has resulted in their widespread acceptance and use in the construction industry. Thus a bar chart is a good tool for communicating construction plans and schedules. Since they are generally used for broad planning, bar charts generally require less revision and updating then the more sophisticated tools.

Limitations of bar charts

They are very cumbersome to use when the number of activities to be represented becomes large. If several sheets are required, logical interconnections between activities are difficult to comprehend. Secondly, logical interconnections between activities are usually not expressed in a bar chart. As a result it becomes quite difficult to reconstruct the logic and to recognize logical constraints unless a substantial amount of documentation is included with the chart. Thirdly, although the bar chart is a good planning and scheduling tool, it is difficult to use it for forecasting the effects that changes in a particular activity will have on the overall schedule or sometimes on the individual activity itself.

12.4.2 Progress Curves (S-curves)

Progress curves graphically represent some measure of cumulative progress on the vertical axis against time on the horizontal axis. The progress can be measured in terms of money spent, work-in-place, man-hours expended etc. The units for these cumulative measures of progress can be absolute units (dollars, cubic metres, etc.) or a percentage of the estimated total quantity. Time, can also be represented either as a percentage of total time or in terms of calendar time.

In a typical project, resource spending starts slowly, builds to a peak and then tapers off near the end of the project. This results in the cumulative curve representing expenditure of a resource to have a relatively small slope at the start, increase during the middle phases of the project, then flatten out towards the end of the project giving the familiar S-shape to the curve. Fig. 12.2 shows how an S-curve can be developed.

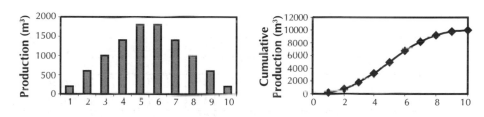

Fig. 12.2 Development of an S-curve

Example 12.2

In this example the use of S-curve in planning, monitoring and controlling is illustrated. In Table 12.2 and Fig. 12.3 the daily production are shown. The first set of data represents *planned* production on a daily basis. The next three sets of data show three different *scenarios* that are possible in reality.

Time, days							Total	
1	2	3	4	5	6	7	Total	
10	20	30	50	35	23	5	175	*Planned*
5	25	35	50	30	20	10	175	*Actual, scenario 1*
10	25	35	55	30	25	10	190	*Actual, scenario 2*
5	15	25	45	25	20	10	145	*Actual, scenario 3*

Table 12.2 Concrete production, daily in cubic metres

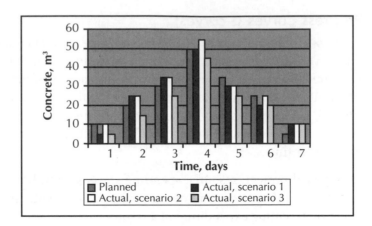

Fig. 12.3 Concrete production, daily in cubic metres

All four sets of cumulative production are tabulated in Table 12.3 and plotted in Fig. 12.4. Note that Fig. 12.4 gives the S-curves for all four sets of data.

Time, days							
1	*2*	*3*	*4*	*5*	*6*	*7*	
10	30	60	110	145	170	175	*Planned*
5	30	65	115	145	165	175	*Actual, scenario 1*
10	35	70	125	155	180	190	*Actual, scenario 2*
5	20	45	90	115	135	145	*Actual, scenario 3*

Table 12.3 Concrete production, cumulative in cubic metres

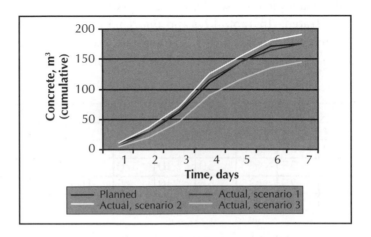

Fig. 12.4 Concrete production, cumulative, in cubic metres (S-curves)

A comparison between "*planned*" and "*actual*, scenario 1" reveals that the actual production progressed more or less as planned without any major deviations. Also, in this scenario, the total actual production in seven days coincided with the total estimated production within the total planned duration of seven days. "*Actual, scenario 2*," however, shows that the actual total production achieved in seven days is 190 cubic metres as opposed to planned amount of 175 cubic metres. Both the "rate of production" as well as the "total production" is underestimated. If the planned production is compared with the "*actual, scenario 3*," we see that the activity is not progressing as planned. The rate of production is lower than expected and at the end of the seven days there is a shortage of 30 cubic metres of concrete production. It is possible that the item was overestimated and further production is not needed. However, it is also possible that more than seven days are needed to finish the task. Thus, if reliable field information is available, S-curves can be used very effectively to plan, monitor and control construction activities and to make related decisions. Project managers often use S-curves for forecasting future performance and for taking corrective actions.

12.4.3 Linear Balance Charts

Linear balance charts or, as they are sometimes called, line of balance (LOB) control charts are best applied to linear and repetitive operations such as tunnels, pipelines, highways and high-rise structures. As shown in Fig. 12.4, the vertical axis typically plots cumulative progress, in terms of units completed/produced or percentage completed for different activities or systems of a project. The horizontal axis plots time. In this example the vertical axis could be in terms of miles or meters of the highway. The sloping lines represent various activities. The project can be considered to be in good shape as long as the slopes of these lines remain parallel to each other as they move to the right. However, if one activity is proceeding too rapidly with a steep slope compared with that of the activity preceding it then time and space conflicts may result. In other words, if one activity is slower, with a flatter slope, compared with its succeeding activities, there will be potential for conflicts.

Example 12.3

An example of a linear balance chart is shown in Fig. 12.4. In this example, the development and use of linear balance chart for planning and monitoring is illustrated. The project involves construction of a highway-segment using concrete pavement. A formwork technique, known as slipforming is used in this project. Slipforming involves moving the formwork as work progresses in a linear fashion resulting in a repetitive sequence of activities. For a simple demonstration of the LOB technique only three major activities are considered in this example. They are:

(I) Cleaning and preparing
(II) Formwork
(III) Pouring concrete

As shown in Fig. 12.4, at the end of the 6th day, 7 units (say, 700 metres) of "cleaning and preparing," and 2 units of formwork are planned. At the end of the 12th day, 15 units of "cleaning and preparing," 11 units of "formwork," and 3 units of "pouring concrete" should be completed. The LOB control chart also shows that between the 6th and 10th day, 6 units of "cleaning and preparing" and 6 units of "formwork" are planned, and during the same time period, no "pouring concrete" are to begin. Fig. 12.4 also includes information on units planned for each two-day period. This type of

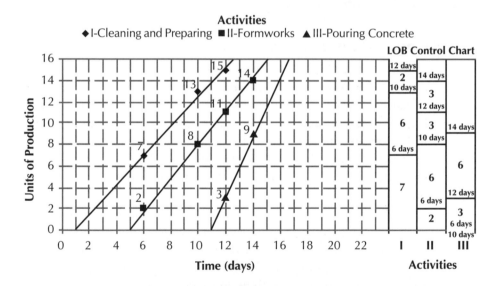

Units planned, periodic										
Days	**2**	**4**	**6**	**8**	**10**	**12**	**14**	**16**	**18**	**20**
I	2	2	3	3	3	2	1	0	0	0
II	0	0	2	3	3	3	3	2	0	0
III	0	0	0	0	0	3	6	6	1	0

Fig. 12.4 Linear balance chart example — Concrete pavement with horizontal slipforming

information is essential to plan procurement and delivery of resources. Actual progress can be directly plotted on the planned LOB chart to detect ensuing problems and to take timely corrective actions. If the actual slopes deviate too much from the planned ones, as pointed out earlier, the project manager should be concerned and take necessary actions.

12.4.4 Network Techniques

With the exception of bar charts, the most common approach to project scheduling is the use of network techniques. There are two network techniques commonly employed in planning and scheduling, i.e. Program Evaluation and Review Technique (PERT) and Critical Path Method (CPM). The PERT was first developed by the United States Navy for the Polaris missile and submarine project in 1958. The CPM was developed by DuPont, Inc., during the same time period (Antill and Woodhead 1990). PERT has been mostly used for research and development projects. CPM was designed for construction projects and has been generally accepted by the construction industry.

The two methods are quite similar in many respects and are often combined for educational presentations. However, originally PERT was oriented more to the time element of projects and used probabilistic time estimates to aid in determining the probability that a project could be completed by some given date. CPM on the other hand, used deterministic activity time estimates and is used to control both the time and cost aspects of a project. Both techniques identify a project critical path whose activities cannot be delayed without increasing the overall project duration.

Networks can be drawn using one of the two techniques: (a) activity on arrow (AOA), and (b) activity on node (AON). The second technique is also known as the precedence method. The details of these two techniques can be found in *Modern Construction Project Management* by Tang, S.L., Poon, S.W., Ahmed S.M. and Wong, K.W., Hong Kong University Press (2003). Most commercially available project management software tools are based on the precedence method since more complex calculations can be performed using this method than the AOA technique.

The major advantages of network techniques over the less sophisticated approaches are:

1. Interdependencies between activities and critical areas are clearly defined. Helps in determining where the greatest effort should be made for a project to stay on schedule.

2. Determination of probability of meeting specified deadlines by the development of alternative plans.
3. Ability to evaluate the effect of changes in the program.
4. Large amount of information can be concisely represented.

The disadvantages of network techniques are:

1. Complexity of network techniques creates difficulties in implementation and limits its use as an effective tool for communication.
2. Expensive to maintain and is generally used on the larger projects.

A simple network is illustrated in Example 12.4 using the precedence (activity on node) technique.

Example 12.4

Consider a list of activities for a water supply project:

No.	Activity	Duration	Predecessor
A	Drill well	4	-
B	Construct power line	3	-
C	Excavate	5	-
D	Deliver material	2	-
E	Pump house	3	A
F	Assemble tank	4	B
G	Foundation	4	C
H	Install pipe	6	C
I	Install pump	2	B, C, E
J	Erect tower & tank	6	F, G

The network diagram with critical path is shown in Fig. 12.5.

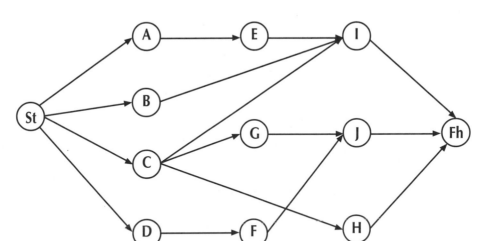

Fig. 12.5 Example network

Decision types a planner/scheduler will face at this point are related to:

a. estimate of durations
b. dependency relationships

Based on the durations and dependencies decided the time-activity computations for this example are as follows. For more detailed descriptions of the technique see Tang, *et al.* (2003).

Act. No.	Description	Duration	Early Start	Early Finish	Late Start	Late Finish	Total Float
A	Drill well	4	0	4	6	10	6
B	Construct power line	2	0	3	3	5	3
C	Excavate	5	0	5	0	5	0
D	Deliver material	3	0	3	10	13	10
E	Pump house	3	4	7	10	13	6
F	Assemble tank	4	2	6	5	9	3
G	Foundation	4	5	9	5	9	0
H	Install pipe	6	5	11	9	15	4
I	Install pump	2	7	9	13	15	6
J	Erect tower & tank	6	9	15	9	15	0

12.5 Cost and Resource Planning

12.5.1 Cost Loading

Cost loading or the distribution of costs, with respect to time, must be considered in order to successfully manage a project. In the preceding sections scheduled early and late starts, and finishes, were calculated based on the duration and sequencing of activities. A cost analysis can also be performed by assigning an anticipated cost to each activity. The cost may be distributed over a range of time, starting from the early to late start and ending from the early to late finish.

Because activities can occur over a range of time, a cost analysis must be performed based on activities starting on an early start, late start, and target schedule. The target schedule can be taken as the midpoint between the early start and the late start during planning. For each day in the project, the cost per day of each activity that is in progress is summed to obtain the total cost of the project for that day. Cumulative project costs are divided by the total project cost to obtain the percentage cost. The percentage time for each day is calculated by dividing the number of the working days by the total project duration. Similar calculations can be performed for activities on a late start schedule and target schedule. S-curves, as explained earlier in this chapter, can be used to represent the cost information.

12.5.2 Resource Loading

The project manager can resource load the project plan to include the number of work-hours required for each activity. The resource plan is similar to the cost distribution analysis presented earlier, except work-hours are used in place of dollars of cost. Thus, the resource plan is simply a histogram of work-hours versus time for each activity or work package.

In a construction project, for example, during the construction phase, the primary resources are labour, materials, and equipment. The correct quantity and quality of material must be ordered and delivered to the job-site at the right time to ensure efficiency of labour. Equipment to be installed in the project often requires a long lead time from the fabricator. Thus, the project plan should include material and equipment required by the construction work force.

The construction plan shows the desired sequence of work. However, to be workable, the plan must also show the distribution of resources, such as the required labour for each craft of the job. The demand for labour should be uniformly distributed for each craft on the project to prevent irregularities. The resource plan can be used as a tool to ensure a relatively uniform distribution of labour on the project.

Exercise Questions

Question 1

"Planning is a decision making process". Explain in the context of a construction project.

Question 2

How does estimating of a construction project involve decision making?

Question 3

Describe the tools and techniques available for scheduling with their relative advantages and disadvantages.

BIBLIOGRAPHY

Chapters 1 and 2

Barzilai, J. (1997). "Deriving weights from pairwise comparison matrices." *Journal of the Operational Research Society*, 48, 1226–1232.

Barzilai, J. and Golany, B. (1994). "AHP rank reversal, normalization and aggregation rules." *INFOR*, 32 (2), 57–64.

Belton V. and Gear, T. (1983). "On a short-coming of Saaty's method of analytic hierarchies." *Omega*, 11 (3), 228–230.

Donegan, H.A., Dodd, F.J. and McMaster, T.B.M. (1992). "A new approach to AHP decision-making." *The Statistician*, 41, 295–302.

Donegan, H.A. Dodd, F.J. and McMaster, T.B.M. (1995). "Theory and methodology — inverse inconsistency in analytic hierarchies." *European Journal of Operational Research*, 80, 86–93.

Johnson, C.R. Beine, W.B. and Wang,T.J. (1979). "Right-left asymmetry in an eigenvector ranking procedure." *Journal of Mathematical Psychology*, 19, 61–64.

Saaty, T.L. (1977). "A scaling method for priorities in hierarchical structures." *Journal of Mathematical Psychology*, 15, 234–281.

Saaty, T.L. (1980). *The Analytic Hierarchy* Process. McGraw-Hill, New York.

Tang, S.L. (1995). "Tender evaluation using Analytic Hierarchy Process." Paper presented in the *1st International Symposium on Project Management*, Northwestern Polytechnical University, Xi'an, China, September 1995.

Tung, S.L. (1997). *Right and left eigenvector inconsistency in analytic hierarchy process.* M.Sc. Dissertation, Civil and Structural Engineering Department, The Hong Kong Polytechnic University.

Tung, S.L. and Tang, S.L. (1998). "A comparison of the Saaty's AHP and Modified AHP for right and left eigenvector inconsistency." *European Journal of Operational Research*, 106, 123–128.

Vargas, L.G. (1982). "Reciprocal matrices with random coefficients." *Mathematical Modelling*, 3, 69–81.

Chapters 3, 4 , 5, 6 and 7

Levin, R.I., Rubin, D.S., Stinson, J.P. and Garden Jr., E.S. (1992). *Quantitative Approaches to Management.* 8th Edition. McGraw-Hill, New York, USA.

Lind, D.A. and Mason, R.D. (1997). *Basic Statistics: for business and economics.* 2nd Edition. Irwin, Times Mirror Higher Education Group, USA.

Meredith, D.D., Wong, K.W., Woodhead, R.W. and Wortman, R.H. (1985). *Design and Planning of Engineering Systems.* 2nd Edition. Prentice-Hall, New Jersey, USA.

Pilcher, R. (1992). *Principles of Construction Management.* 3rd Edition. McGraw-Hill, London, U.K.

Tang, S.L. and Poon, S.W. (1987). *Project Management, Vol. 2.* Education Technology Unit, Hong Kong Polytechnic, Hong Kong.

Chapter 8

Bather, J. (2000). *Decision Theory: An Introduction to Dynamic Programming and Sequential Decisions.* John Wiley & Sons, New York, USA.

Bellman, R.E. (1957). *Dynamic Programming.* Princeton University Press, New Jersey, USA.

Jackson, W. (1995). *Optimization.* University of London, London, U.K.

Makower M.S. and Williamson, E. (1985). *Operational Research*. 4th Edition. Hodder and Stoughton Education, UK.

Ozan, T.M. (1986). *Applied Mathematical Programming for Engineering and Production Management*. Prentice-Hall, New Jersey, USA.

Chapters 9 and 10

AbouRizk, S. M. and Halpin, D. W. (1991). "Visual Interactive Fitting of Beta Distributions" *Journal of Construction Engineering and Management*, ASCE, 117 (4), 589–605.

AbouRizk, S. M. and Halpin, D. W. (1994). "Fitting beta distributions based on sample data." *Journal of Construction Engineering and Management*, ASCE, 120 (2), 288–305.

AbouRizk, S.M. (2000). *Simphony CYCLONE user's guide*, Construction Engineering & Management Program, University of Alberta, Canada.

Fente, J., Schexnayder, C. and Knutson, K. (2000) "Defining a probability distribution function for construction simulation." *Journal of Construction Engineering and Management*, ASCE, 126 (3), 234–241.

Lichtenstein, S., Fischhoff, B. and Phillips, L.D. (1977). "Calibration of probabilities: the state of the art to 1980." *Judgment under uncertainty: heuristics and biases*. Cambridge University Press, U.K. 306-334.

Halpin, D.W. (1977). "CYCLONE-A method for modeling job site processes." *Journal of Construction Division*, ASCE, 103 (3), 489–499.

Halpin, D.W. and Riggs, L. (1992). *Planning and analysis of construction operations*. Wiley, New York, USA.

Law A. and Kelton D. (1982). *Simulation Modeling and Analysis*. McGraw Hill, New York, USA.

Lu, M. (2002). "Enhancing PERT simulation through ANN-based input modeling." *Journal of Construction Engineering and Management*, ASCE, 128 (5), 438–445.

Lu, M., and Anson, M. (2004). "Establish concrete placing rates using quality control records from Hong Kong building construction projects." *Journal of Construction Engineering and Management*, ASCE. Vol. 130, March/April Issue.

Pidd, M. (1989). *Computer Simulation in Management Sciences*. 2nd ed. Wiley, New York, USA.

Chapters 11 and 12

Kroenke, D., and Hatch, R. (1994). *Management of Information Systems*, 3rd edition, McGraw-Hill, Singapore.

Tenah, K. A. (1984). "Management of Information in Organizations and Routing". *Journal of Construction Engineering and Management*, ASCE, 110(1), pp. 101–118.

Antill, J.M. and Woodhead, R.W. (1990). *Critical Path Methods in Construction Practice*. 4th ed., Wiley, NY.

Tang, S.L., Poon, S.W., Ahmed, S.M. and Wong, K.W. (2003). *Modern Construction Project Management*, 2nd ed., Hong Kong University Press, Hong Kong.

Answers to Selected Exercise Questions

Chapter 1

Question 2:

It is important to make sure that the levels are totally independent, and it should not be possible to merge any of the two or more levels. Here are two examples of 5-Level hierarchy structures to evaluate construction engineering decision problems:

Solution 1

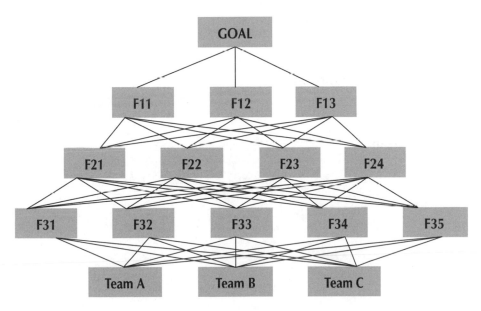

Level 1

Goal: to determine the most suitable team for this project.

Level 2

Components of the Project

F11: Departmental Office
F12: Residential Hall
F13: Scientific Laboratory

Level 3

Individual component to be designed in each component

F21: Air-conditioning system
F22: Lighting system
F23: Fire resisting system
F24: Piping system

Level 4

Factors affecting the works

F31: Team's experience
F32: Team's professional skills
F33: Duration for completing the project
F34: Identifying key issues
F35: Innovative idea

Level 5

Team alternatives

1: Team A
2: Team B
3: Team C

Solution 2

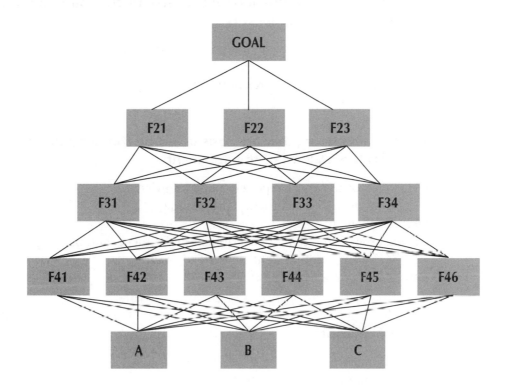

Level 1

Goal: to rank the 3 applicants.

Level 2

The 3 Assessors

F21: Project manager from the client
F22: Project director
F23: Project engineer from the consultant

Level 3

The 3 sources of information for the recruitment board to evaluate the abilities of the applicants

F31: Application package submitted by the applicant
F32: The interviews
F33: Past performance records of the applicants working for the client

Level 4

The 6 main factors that are considered by the recruitment board to assess the abilities of the applicants

F41: Relevant technical experience and knowledge
F42: Monitoring of construction programme
F43: Resolving of contractual issues
F44: Liaison with different government departments and other agencies
F45: Management of resident site staff
F46: Management of the contractor

Level 5

The 3 applicants:

1: Applicant A
2: Applicant B
3: Applicant C

Chapter 2

Question 1:

Belton and Gear used the following to demonstrate their famous examples of rank reversal:

Example 1

Suppose that the decision maker wishes to evaluate three options (A, B and C) on three criteria (a, b and c). Saaty's approach is outlined in the following steps:

Step 1: by comparing options on each criterion, the following matrices (consistent) are formed:

Criterion a   Criterion b   Criterion c

$$\begin{pmatrix} 1 & 1/9 & 1 \\ 9 & 1 & 9 \\ 1 & 1/9 & 1 \end{pmatrix} \quad \begin{pmatrix} 1 & 9 & 9 \\ 1/9 & 1 & 1 \\ 1/9 & 1 & 1 \end{pmatrix} \quad \begin{pmatrix} 1 & 8/9 & 8 \\ 8/9 & 1 & 9 \\ 1/8 & 1/9 & 1 \end{pmatrix}$$

Step 2: calculate relative weightings of options on each criterion

It can be shown that the normalised priority vectors are:

Criterion   Criterion   Criterion
a   b   c
(W1)   (W2)   (W3)

$$\begin{pmatrix} 1/11 \\ 9/11 \\ 1/11 \end{pmatrix} \quad \begin{pmatrix} 9/11 \\ 1/11 \\ 1/11 \end{pmatrix} \quad \begin{pmatrix} 8/18 \\ 9/18 \\ 1/18 \end{pmatrix}$$

The weighting matrix (or composite matrix) W = (W1, W2, W3)

Step 3:

For the comparison of Level 2 with respect to Level 1, the three criteria are assumed to be of equal importance. The matrix therefore is $(1/3, 1/3, 1/3)^1$.

Step 4:

The overall importance of each option is obtained by multiplying the composite matrix in Step 2 by the matrix in Step 3. The overall priority vector is (0.45, 0.47, 0.08). i.e. B > A > C.

Example 2

Now suppose that there is an additional alternative D, which is exactly similar to B, is introduced to the problem and that the judgement matrices of Step 1 become:

Criterion a   *Criterion b*   *Criterion c*

$$
\begin{pmatrix} 1 & 1/9 & 1 & 1/9 \\ 9 & 1 & 9 & 1 \\ 1 & 1/9 & 1 & 1/9 \\ 9 & 1 & 9 & 1 \end{pmatrix}
\quad
\begin{pmatrix} 1 & 9 & 9 & 9 \\ 1/9 & 1 & 1 & 1 \\ 1/9 & 1 & 1 & 1 \\ 1/9 & 1 & 1 & 1 \end{pmatrix}
\quad
\begin{pmatrix} 1 & 8/9 & 8 & 8/9 \\ 9/8 & 1 & 9 & 1 \\ 1/8 & 1/9 & 1 & 1/9 \\ 9/8 & 1 & 9 & 1 \end{pmatrix}
$$

with corresonding eigenvectors (Step 2)

Criterion a   *Criterion b*   *Criterion c*

$$
\begin{pmatrix} 1/20 \\ 9/20 \\ 1/20 \\ 9/20 \end{pmatrix}
\quad
\begin{pmatrix} 9/12 \\ 1/12 \\ 1/12 \\ 1/12 \end{pmatrix}
\quad
\begin{pmatrix} 8/27 \\ 9/27 \\ 1/27 \\ 9/27 \end{pmatrix}
$$

Since the criteria are still assumed to be of equal importance compared with respective to Level 1 so the matrix is again $(1/3, 1/3, 1/3)^{\mathrm{T}}$. The overall priority vector of the options is therefore calculated to be $(0.37, 0.29, 0.06, 0.29)$, i.e. $A > B \sim D > C$.

It can be observed that the rankings of Examples 1 and 2 are not consistent: the ranking of Options A and B is reversed after the introduction of Option D, although the pairwise inputs connected with A, B and C options (and D of course) are unchanged.

This is one of the most famous examples used to demonstrate the problem of rank reversal, frequently referred to by other researchers as the Belton and Gear's rank reversal.

Question 2:

In their paper, Barzilai and Golany analysed the Belton and Gear's rank reversal with Geometric Mean method. They showed that rank reversal could be avoided if the geometric mean and the weighted-geometric-mean aggregation rule are used to tackle the problem. It was claimed that this method worked because it preserved the underlying mathematical structure.

For more detail, one could read Barzilai and Golany's paper; but in short, this was how Barzilai and Golany approached the Belton and Gear rank reversal problem:

Instead of finding the priority vector, the original criterioa a, b and c are used and their geometric means are calculated:

Geometric Mean:

Example 1 in Question 1 with geometric mean:

$$U^* = \begin{pmatrix} 1 & 1/9 & 1 \\ 9 & 1 & 9 \\ 1 & 1/9 & 1 \end{pmatrix}^{1/3} \times \begin{pmatrix} 1 & 9 & 9 \\ 1/9 & 1 & 1 \\ 1/9 & 1 & 1 \end{pmatrix}^{1/3} \times \begin{pmatrix} 1 & 8/9 & 8 \\ 8/9 & 1 & 9 \\ 1/8 & 1/9 & 1 \end{pmatrix}^{1/3}$$

$$= \begin{pmatrix} 1 & 0.961 & 4.160 \\ 1.040 & 1 & 4.327 \\ 0.240 & 0.231 & 1 \end{pmatrix}$$

$$Priortiy\ Vector = \begin{pmatrix} (1 & 0.961 & 4.160)^{1/3} \\ (1.040 & 1 & 4.327)^{1/3} \\ (0.240 & 0.231 & 1)^{1/3} \end{pmatrix} = \begin{pmatrix} 1.587 \\ 1.651 \\ 0.381 \end{pmatrix}$$

i.e. $B > A > C$

Example 2 in Question 1 with geometric mean:

$$U = \begin{pmatrix} 1 & 1/9 & 1 & 1/9 \\ 9 & 1 & 9 & 1 \\ 1 & 1/9 & 1 & 1/9 \\ 9 & 1 & 9 & 1 \end{pmatrix}^{1/3} \times \begin{pmatrix} 1 & 9 & 9 & 9 \\ 1/9 & 1 & 1 & 1 \\ 1/9 & 1 & 1 & 1 \\ 1/9 & 1 & 1 & 1 \end{pmatrix}^{1/3} \times \begin{pmatrix} 1 & 8/9 & 8 & 8/9 \\ 9/8 & 1 & 9 & 1 \\ 1/8 & 1/9 & 1 & 1/9 \\ 9/8 & 1 & 9 & 1 \end{pmatrix}^{1/3}$$

$$= \begin{pmatrix} 1 & 0.961 & 4.160 & 0.961 \\ 1.040 & 1 & 4.327 & 1 \\ 0.240 & 0.231 & 1 & 0.231 \\ 1.040 & 1 & 4.327 & 1 \end{pmatrix}$$

$$Priortiy\ Vector = \begin{pmatrix} (1 & 0.961 & 4.160 & 0.961)^{1/3} \\ (1.040 & 1 & 4.327 & 1)^{1/3} \\ (0.240 & 0.231 & 1 & 0.231)^{1/3} \\ (1.040 & 1 & 4.327 & 1)^{1/3} \end{pmatrix} = \begin{pmatrix} 1.621 \\ 1.651 \\ 0.234 \\ 1.651 \end{pmatrix}$$

i.e $B \sim D > A > C$

SAHP

The ranking according to SAHP is already shown in Question 1 answer, that is, $B > A > C$ for Example 1 and $A > B \sim D > C$ for Example 2.

MAHP:

First, 8/9 and 9/8 from the criterion c matrix must be mapped manually with the method stated in Section 2.1 of Chapter 2.

8/9 —> 0.9877   9/8 —> 1.014

The modified matrices are as below:
Example 1 with MAHP:

$$
\begin{array}{ccc}
\textit{Criterion a} & \textit{Criterion b} & \textit{Criterion c}
\end{array}
$$

$$
\begin{pmatrix} 1 & 0.243 & 1 \\ 4.123 & 1 & 4.123 \\ 1 & 0.243 & 1 \end{pmatrix}
\begin{pmatrix} 1 & 4.123 & 4.123 \\ 0.243 & 1 & 1 \\ 0.243 & 1 & 1 \end{pmatrix}
\begin{pmatrix} 1 & 0.988 & 2.828 \\ 1.014 & 1 & 4.123 \\ 0.354 & 0.243 & 1 \end{pmatrix}
$$

$$
\textit{Priortiy Vector} =
\begin{pmatrix} 0.16 & 0.67 & 0.41 \\ 0.67 & 0.16 & 0.47 \\ 0.16 & 0.16 & 0.13 \end{pmatrix}
\times
\begin{pmatrix} 1/3 \\ 1/3 \\ 1/3 \end{pmatrix}
=
\begin{pmatrix} 0.41 \\ 0.43 \\ 0.15 \end{pmatrix}
$$

Overall ranking = B > A > C

Example 2 with MAHP:

$$
\begin{array}{cc}
\textit{Criterion a} & \textit{Criterion b}
\end{array}
$$

$$
\begin{pmatrix} 1 & 0.243 & 1 & 0.243 \\ 4.123 & 1 & 4.123 & 1 \\ 1 & 0.243 & 1 & 0.243 \\ 4.123 & 1 & 4.123 & 1 \end{pmatrix}
\begin{pmatrix} 1 & 4.123 & 4.123 & 4.123 \\ 0.243 & 1 & 1 & 1 \\ 0.243 & 1 & 1 & 1 \\ 0.243 & 1 & 1 & 1 \end{pmatrix}
$$

$$
\textit{Criterion c}
$$

$$
\begin{pmatrix} 1 & 0.988 & 2.828 & 0.988 \\ 1.014 & 1 & 4.123 & 1 \\ 0.354 & 0.243 & 1 & 0.243 \\ 1.014 & 1 & 4.123 & 1 \end{pmatrix}
$$

$$
\textit{Priortiy Vector} =
\begin{pmatrix} 0.10 & 0.58 & 0.29 \\ 0.40 & 0.14 & 0.32 \\ 0.10 & 0.14 & 0.08 \\ 0.40 & 0.14 & 0.32 \end{pmatrix}
\times
\begin{pmatrix} 1/3 \\ 1/3 \\ 1/3 \end{pmatrix}
=
\begin{pmatrix} 0.32 \\ 0.29 \\ 0.11 \\ 0.29 \end{pmatrix}
$$

Overall ranking = A > B~D > C
Rank reversed

Chapter 3

Question 1:

The decision tree for the problem is:

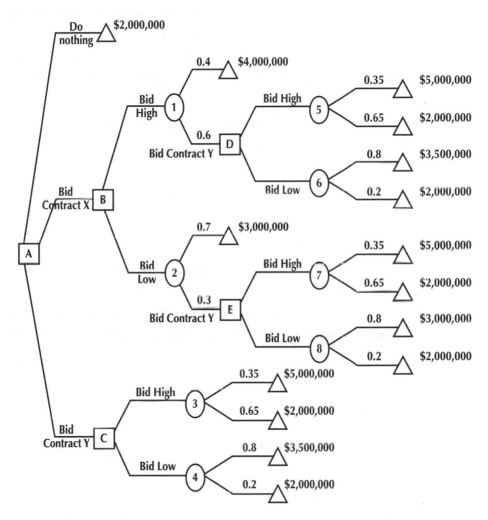

The EMVs at chance nodes 3, 4, 5, 6, 7 and 8 are calculated below:

$$EMV_3 = 0.35 \times 5,000,000 + 0.65 \times 2,000,000$$
$$= \$3,050,000$$

$$EMV^4 = 0.8 \times 3,500,000 + 0.2 \times 2,000,000$$
$$= \$3,200,000$$

$EMV_5 = EMV_3 = \$3,050,000$
$EMV_6 = EMV_4 = \$3,200,000$
$EMV_7 = EMV_3 = \$3,050,000$
$EMV_8 = EMV_4 = \$3,200,000$

It can be seen that we have to make decisions at decision nodes C, D and E before the EMVs at chance node 1 and 2 can be calculated.

Therefore, at decision node C, select EMV_4 ($\$3,200,000$); at decision node D, select EMV_6 ($\$3,200,000$) and at decision node E, select EMV_8 ($\$3,200,000$).

Now we can calculate the EMVs at chance nodes 1 and 2.

EMV_1 = $0.4 \times 4,000,000 + 0.6 \times 3,200,000$ (EMV_6)
= $\$3,520,000$

EMV_2 = $0.7 \times 3,000,000 + 0.3 \times 3,200,000$ (EMV_8)
= $\$3,060,000$

At decision node B, select EMV_1 ($\$3,520,000$).

Finally, at decision node A, select EMV_1 (highest of $\$2,000,000$, $\$3,520,000$ and $\$3,200,000$).

Therefore, the decision should be to bid for contract X and submit high tender price, and if unsuccessful, then bid for contract Y and submit low tender price.

Chapter 4

Question 1:

(a). The decision tree (based on monetary values) is:

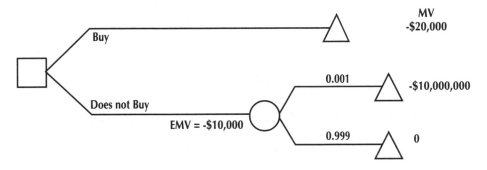

Therefore, according to that EMV criterion, the contractor should not buy the insurance.

(b). Assign utility values of 1,000 to $0 and 0 to - $10,000,000.

The utility value of - $20,000 is found by making use of the following hypothetical problem.

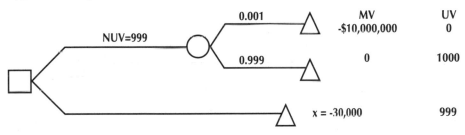

The utility curve is:

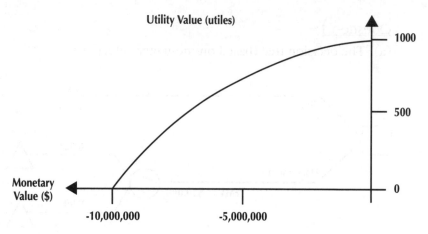

The utility values are found from the above utility curve:

Monetary value	Utility value
-30,000	999
-20,000	999.3

The decision tree (based on utility values) becomes:

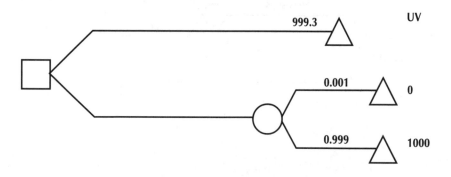

Therefore, according to the EUV criterion, the contractor should buy the insurance.

Chapter 5

Question 2:

We start by forming the Basic Decision Tree for the repairing machine problem.

The Basic Decision Tree of the problem is:

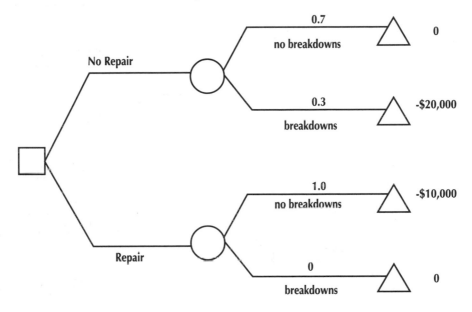

Consider 2 types of test alternatives:

1. No Test.
2. Carry out a dynamometer test for the condition of the machine to determine whether a repair is required.

The dynamometer test is an imperfect test with a sample likelihood as follows:

	T_1 (Good Condition)	T_2 (Bad Condition)
S_1 (Predict no Breakdowns [Good condition])	0.85	0.15
S_2 (Predict Breakdowns [Bad Condition])	0.15	0.85
	$\Sigma = 1.00$	$\Sigma = 1.00$

The decision tree with conditional probabilities if dynamometer test is performed:

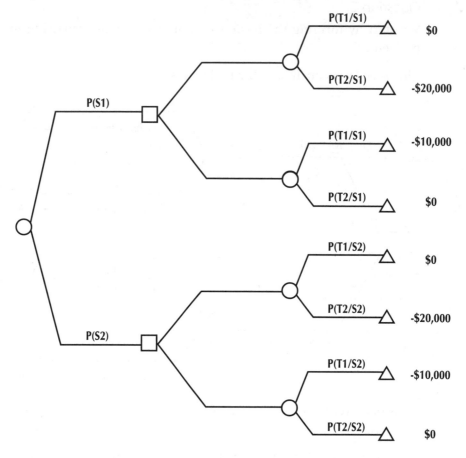

$P(S_i) =$ probability of occurrence of S_i under the given sample likelihood with the assumed probabilities of 70:30 chance of the machine not breaking down.

$P(T_j/S_i) =$ probability of occurrence of T_j when predicted breakdowns S_i occurs.

where i = 1 or 2 and j = 1 or 2

The Bayes' Theorem states that: $P(T_j / S_i) = \dfrac{P(S_i/T_j) \times P(T_j)}{P(S_i)}$

Now, the values of $P(S_i)$ should be computed as follows:

$$P(T_1) = 0.7 \quad \begin{array}{l} P(S_1/T_1) = 0.85 \\ P(S_2/T_1) = 0.15 \end{array}$$
$$P(T_2) = 0.3 \quad \begin{array}{l} P(S_1/T_2) = 0.15 \\ P(S_2/T_2) = 0.85 \end{array}$$

0.7 x 0.85 = 0.595, i.e. $P(S_1 \wedge T_1)$

0.7 x 0.15 = 0.105, i.e. $P(S_2 \wedge T_1)$

0.3 x 0.15 = 0.045, i.e. $P(S_1 \wedge T_2)$

0.3 x 0.85 = 0.255, i.e. $P(S_2 \wedge T_2)$

Since $P(S_i) = P(S_i \wedge T_1) + P(S_i \wedge T_2)$

Hence,

$$P(S_1) = P(S_1 \wedge T_1) + P(S_1 \wedge T_2) = 0.595 + 0.045 = 0.640$$
$$P(S_2) = P(S_2 \wedge T_1) + P(S_2 \wedge T_2) = 0.105 + 0.255 = 0.360$$

The values of $P(T_j/S_i)$ in our problem can also be computed as follows:

$$P(T_1/S_1) = \frac{P(S_1/T_1) \times P(T_1)}{P(S_1)} = \frac{0.85 \times 0.7}{0.640} = 0.930$$

$$P(T_2/S_1) = \frac{P(S_1/T_2) \times P(T_2)}{P(S_1)} = \frac{0.15 \times 0.3}{0.640} = 0.070$$

$$P(T_1/S_2) = \frac{P(S_2/T_1) \times P(T_1)}{P(S_2)} = \frac{0.15 \times 0.7}{0.360} = 0.292$$

$$P(T_2/S_2) = \frac{P(S_2/T_2) \times P(T_2)}{P(S_2)} = \frac{0.85 \times 0.3}{0.360} = 0.708$$

The overall decision tree for the problem:

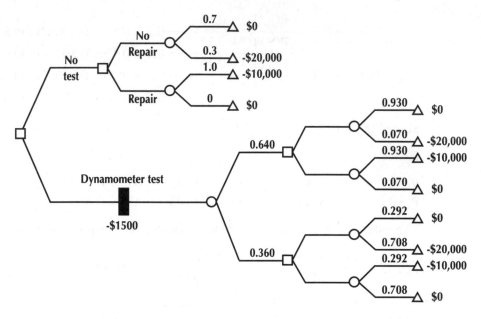

The overall EMV for this problem:

(1). EMV for no test is -$6000

(2). EMV for dynamometer test is -$3447.20.

Therefore, the project manager should carry out the dynamometer test then decide whether the machine should be repaired based on the result of the dynamometer test.

Chapter 6

Question 2:

$$T = \text{ordering cost} + \text{carrying cost} = \frac{AP}{x} + \frac{xRC}{2}(1 - \frac{u}{r})$$

$$\frac{dT}{dx} = -\frac{AP}{x^2} + \frac{RC}{2}(1 - \frac{u}{r})$$

For minimum T:

$$\frac{dT}{dx} = 0 \quad i.e. \frac{AP}{x^2} = \frac{RC}{2}(1 - \frac{u}{r})$$

$$So, \frac{AP}{x} = \frac{xRC}{2}(1 - \frac{u}{r})$$

i.e. ordering cost = carrying cost

Question 3:

(a) Ordering cost in a year = P × number of orders in a year = $P\frac{A}{Q}$

Carrying cost in a year = $l_{ave} \times C\% = \frac{RQ}{2}\frac{C}{100}$

$$T = \frac{\text{ordering cost}}{\text{year}} + \frac{\text{carrying cost}}{\text{year}} = \frac{PA}{Q} + \frac{CRQ}{200}$$

To find the minimum T, let:

$$\frac{dT}{dx} = 0 \quad So, -\frac{PA}{Q^2} + \frac{CR}{200} = 0 \quad i.e. \ Q^2 = \frac{200AP}{RC}$$

(b) P = 500
   A = 1,000,000
   R = 100
   C = 10%

Since $Q^2 = \frac{200AP}{RC} = \frac{200 \times 1,000,000 \times 500}{100 \times 10} = 100,000,000$

$$\therefore Q = 10,000$$

Therefore, 10,000 tonnes of cement should be ordered each time and there should be 100 orders (1,000,000 / 10,000) in a year.

(c) If   P = 1000
          C = 5%

$$\text{Then } Q = \sqrt{\frac{200 \times 1,000,000 \times 1000}{100 \times 5}} = 20,000$$

For Q = 20,000
(From Part a.)

$$T = \frac{PA}{Q} + \frac{CRQ}{200} = \frac{1000 \times 1,000,000}{20,000} + \frac{5 \times 100 \times 20,000}{200}$$

$$= 50,000 + 50,000$$
$$= \$100,000$$

For Q = 10,000

$$T = \frac{1000 \times 1,000,000}{10,000} + \frac{5 \times 100 \times 10,000}{200}$$

$$= 100,000 + 25\ 000$$
$$= \$125,000$$

Therefore, loss in the year due to incorrect estimation

$$= \$125,000 - \$100,000$$
$$= \underline{\$25,000}$$

Chapter 7

<u>Question 1</u>:

        h = carrying cost / pipe / week
        s = stocking cost / pipe / week
        P = ordering cost / order
        u = usage rate (number of pipe / week)

Now,   h = $500 × 5% = $25
        s = $100
        P = $100
        u = 500

Therefore EOQ = y

$$= \sqrt{\frac{2Pu}{h}} \ \sqrt{\frac{h+s}{s}} = \sqrt{\frac{2 \times 100 \times 500}{25}} \ \sqrt{\frac{25+100}{100}} = 71$$

Frequency of delivery (or frequency of batch)

$$= \frac{71}{500} \ week = \frac{71 \times 7}{500} \ days = 1 \ day$$

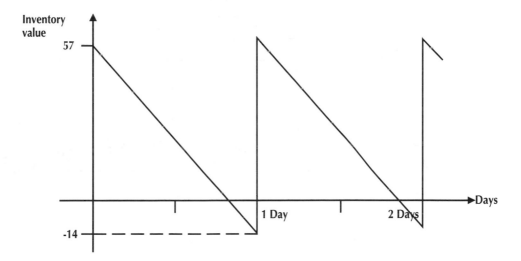

Chapter 8

Question 2:

Stages — 4 Stations (n = 1, 2, 3, 4)
n = Number of Stations (1, 2, 3, 4)
Stage S = Number of inspectors avaliable at station n
X_n = Number of inspectors assigned to station n

Return Function $\quad f_n (S, X_n) = P_n (X_n) \times F_{n-1}{}^* (S - X_n)$

Revise Formula $\quad f_n{}^* (S) = \min \{f_n (S, X_n)\}$
$\qquad\qquad\qquad = \min \{P_n (X_n) \times F_{n-1}{}^* (S - X_n)\}$

Stage 1   *Station 4 only*

S \ X_1	$f_1(S, X_2) = P_1(X_1)$			$f_1{}^*(S)$	$X_1{}^*$
	1	2	3		
1	0.2			0.2	1
2	0.2	0.1		0.1	2
3	0.2	0.1	0.1	0.1	2, 3

Stage 2   *Station 3 AND Station 4*

S \ X_2	$f_2(S, X_2) = P_2(X_2) \times f_1{}^*(S - X_2)$			$f_2{}^*(S)$	$X_2{}^*$
	1	2	3		
2	$0.3 \times 0.2 = 0.06$			0.06	1
3	$0.3 \times 0.1 = 0.03$	$0.15 \times 0.2 = 0.03$		0.03	1,2
4	$0.3 \times 0.1 = 0.03$	$0.15 \times 0.1 = 0.015$	$0.05 \times 0.2 = 0.01$	0.01	3

Stage 3   *Station 2 AND (Station 3 AND Station 4)*

S \ X_3	$f_3(S, X_3) = P_3(X_3) \times f_2^*(S - X_3)$			$f_3^*(S)$	X_3^*
	1	2	3		
3	$0.4 \times 0.06 =$ 0.024			0.024	1
4	$0.4 \times 0.03 =$ 0.012	$0.3 \times 0.06 =$ 0.018		0.012	1
5	$0.4 \times 0.01 =$ 0.004	$0.3 \times 0.03 =$ 0.009	$0.1 \times 0.06 =$ 0.006	0.004	1

Stage 4   *Station 1 AND (Station 2 AND (Station 3 AND Station4))*

S \ X_4	$f_4(S, X_4) = P_4(X_4) \times f_3^*(S - X_4)$			$f_4^*(S)$	X_4^*
	1	2	3		
4	$0.3 \times 0.024 =$ 0.0072			0.0072	1
5	$0.3 \times 0.012 =$ 0.0036	$0.2 \times 0.024 =$ 0.0048		0.0036	1
6	$0.3 \times 0.004 =$ 0.0012	$0.2 \times 0.012 =$ 0.0024	$0.1 \times 0.024 =$ 0.0024	0.0012	1

Solution:

Station	No. of inspectors
1	1
2	1
3	3
4	1
total =	6

Chapter 9

Question 1:

Step 1

Assign Random numbers (R.N.) for size of B's order:

Size of B's order	P	Cumulative P	R.N.
5	0.2	0.2	0-19
7	0.3	0.5	20-49
9	0.3	0.8	50-79
11	0.2	1.0	80-99

Size of the B's order from the random numbers generated:

DAY	5	10	15	20	25	30
R.N.	75	39	97	11	4	24
Size of B's Order	9	7	11	5	5	7

Step 2

Assign random numbers for days between C's order:

Days between C's order	P	Cumulative P	R.N.
3	0.1	0.1	0-9
4	0.4	0.5	10-49
5	0.3	0.8	50-79
6	0.2	1.0	80-99

Days between C's order from the random numbers generated:

R.N.	85	39	97	11	4	24	54	3	88	99	57
Days between C's order	6	3	6	4	3	4	5	3	6	6	5

Step 3

Assign random numbers for size of D's Order:

Size of D's order (units)	Probability	Cumulative Probability	R.N.
3	0.1	0.1	0-9
6	0.2	0.3	10-29
9	0.4	0.7	30-69
12	0.3	1.0	70-99

Size of the D's order from the random numbers generated:

R.N.	42	3	81	66	30	52	21	16	4	47
Size of D's Order	9	3	12	9	9	9	6	6	3	9

Assign random numbers for days between D's order:

Days between D's order	Probability	Cumulative Probability	R.N.
2	0.2	0.2	0-19
3	0.3	0.5	20-49
4	0.3	0.8	50-79
5	0.2	1.0	80-99

Days between D's order from the random numbers generated:

R.N.	87	56	56	12	33	70	42	53	21	60	87
Days between D's order	5	4	4	2	3	4	3	4	3	4	5

Step 4

Draw simulation result for 25 days demand for concrete:

Plant / Day	A	B	C	D
1	7	9	5	9
2				
3				
4				
5	7			
6		7		3
7			5	
8				
9	7			
10			5	12
11		11		
12				
13	7			
14				9
15				
16		5	5	9
17	7			
18				
19				9
20			5	
21	7	5		
22				
23			5	6
24				
25	7			
Sub. Total	49	37	30	57

From the simulation, total concrete ordered by the three contractors A,B,C and D in 25 days is (49+37+30+57) = 173.

Hence the estimated mean demand per day for concrete = (173/25) = 6.92 units.

Question 3:

Step 1

Find the value of factor K for each year, K = D/Q:

Year	Discharge, Q (m³ /sec)	Damage, D ($10⁶)	K
1993	6.50	2.00	0.308
1994	9.60	3.40	0.354
1995	12.50	5.00	0.400
1996	3.70	0.80	0.216
1997	5.70	1.50	0.263
1998	17.00	6.00	0.353
1999	4.80	1.50	0.313
2000	9.90	2.80	0.283
2001	12.10	5.50	0.455
2002	8.50	3.00	0.353

Step 2

Assign Random Numbers (R.N.) for Discharge rate, Q:

Discharge, Q (m³ / sec)	Probability	Cumulative Probability	R.N.
6.50	0.10	0.10	0-9
9.60	0.10	0.20	10-19
12.50	0.10	0.30	20-29
3.70	0.10	0.40	30-39
5.70	0.10	0.50	40-49
17.00	0.10	0.60	50-59
4.80	0.10	0.70	60-69
9.90	0.10	0.80	70-79
12.10	0.10	0.90	80-89
8.50	0.10	1.00	90-99
	1.00		

Rate of discharge, Q from the random numbers generated:

Year	R.N.	Discharge, Q (m³/sec)	Year	R.N.	Discharge, Q (m³/sec)
1	16	9.6	14	7	6.5
2	83	12.1	15	79	9.9
3	4	6.5	16	74	9.9
4	66	4.8	17	13	9.6
5	18	9.6	18	37	3.7
6	37	3.7	19	92	8.5
7	30	3.7	20	97	8.5
8	53	17	21	39	3.7
9	63	4.8	22	46	5.7
10	33	3.7	23	94	8.5
11	24	12.5	24	56	17
12	73	9.9	25	29	12.5
13	47	5.7			

Step 3

Assign the random numbers for factor K:

K	Probability	Cumulative Probability	R.N.
0.308	0.10	0.10	0-9
0.354	0.10	0.20	10-19
0.400	0.10	0.30	20-29
0.216	0.10	0.40	30-39
0.263	0.10	0.50	40-49
0.313	0.10	0.60	50-59
0.283	0.10	0.70	60-69
0.455	0.10	0.80	70-79
0.353	0.20	1.00	80-99
	0.10		

The value of the multiplication factor, K from the random numbers generated:

Year	R.N.	K
1	91	0.353
2	9	0.308
3	82	0.353
4	30	0.216
5	21	0.400
6	2	0.308
7	67	0.283
8	79	0.455
9	25	0.400
10	0	0.308
11	84	0.353
12	92	0.353
13	62	0.283
14	87	0.353
15	61	0.283
16	30	0.216
17	94	0.353
18	70	0.455
19	99	0.353
20	9	0.308
21	19	0.354
22	70	0.455
23	40	0.263
24	95	0.353
25	63	0.283

Step 4

With the data obtained, calculate the damage for each year, $D = K \times Q$, for 25 years:

	Discharge, Q (m3 /sec)	K	Damage, D ($106)
1	9.6	0.353	$3.39
2	12.1	0.308	$3.73
3	6.5	0.353	$2.29
4	4.8	0.216	$1.04
5	9.6	0.400	$3.84
6	3.7	0.308	$1.14
7	3.7	0.283	$1.05
8	17	0.455	$7.74
9	4.8	0.400	$1.92
10	3.7	0.308	$1.14
11	12.5	0.353	$4.41
12	9.9	0.353	$3.49
13	5.7	0.283	$1.61
14	6.5	0.353	$2.29
15	9.9	0.283	$2.80
16	9.9	0.216	$2.14
17	9.6	0.353	$3.39
18	3.7	0.455	$1.68
19	8.5	0.353	$3.00
20	8.5	0.308	$2.62
21	3.7	0.354	$1.31
22	5.7	0.455	$2.59
23	8.5	0.263	$2.24
24	17	0.353	$6.00
25	12.5	0.283	$3.54
			$70.39

From the table the anticipated total flood damage for the next 25 years is 70.39×10^6 .

Chapter 10

Question 2:

The figure is a CYCLONE model that describes the operation of a road pavement maintenance project. It assumes there are initially 4 asphalt trucks ready to supply hot asphalt mix for the wear-course paving operation, i.e. four resource entities are initialized at the "Trucks of Asphalt" Que node. To initiate the "Park before Spread" COMBIN activity, an information unit called "Request Signal" as initialized at the "Request Signal" Que node, should also be available, denoting the parking space before the spreader is vacant. Once the a asphalt truck is processed at the "Park before Spreader" COMBI activity, five resource entities are created and held at the "Truck to Paver's Hopper" Que node, each representing the asphalt mix on one paving section. The next step is the "Unload to Paver" COMBI activity, which requires the asphalt mix for one paving section together with an ready (or idle) paver machine to start. Note only one paver is used in the system and is initialized at the "Paver Ready" Que node. Following the "Unloading to Paver" activity is a CONSOLIDATE function node "CON 5" used to count the number of paving sections completed, and a "Paving" NOMAL activity for representing the time delay of the paving operation on one paving section. Also note that once five entities are accumulated at the "CON 5" function node, meaning once five paving sections are completed, an entity will be fired to model an emptied asphalt truck, which then goes through the "Truck Leaving" NOMAL activity. After the time delay over the "Truck Leave" activity, an information unit returns to the "Request Signal" Que node to invoke the processing of the next asphalt truck. On the other hand, subsequent to the "Paving" NORMAL activity, the paver returns to the "Paver Ready" Que node, ready for paving the next section, while another CONSOLIDATE function node "CON 2" succeeding the "Paving" activity is used to accumulate two paving sections and then transform them into one compaction section for further compaction processing (Note that the length of one compaction section is twice that of a paving section as stated in the paving method). In the model, a resource entity representing a ready compaction section enters the "Compact Section" Que node, which should be set as zero at the start (i.e. no compaction section is ready for processing at the start). To process one compaction section by the "Compact Pavement" COMBI activity, an idle compact roller is also needed, which is modeled with the "Compact Roller" Que node. The COUNTER node "10 hits to stop" ensuing the "Compact Pavement" COMBI activity records the number of compaction sections completed; once ten sections are completed, the paving job is done and the simulation terminated.

Chapter 11

Question 1:

A business organization cannot be effective without its information system being effective. Thus organizational structures or hierarchies must serve the business's information processing needs. The structure or the hierarchy should be designed in order to facilitate effective information processing. One should be compatible with the other. A flat organizational structure, for example, with less of a hierarchy, would be suitable for an organization where a formal chain of command is not essential for carrying out its functions. A vertical structure, on the other hand, would be needed in an organization where a rigid and formal chain of command is necessary. Thus the way information is processed and decisions are made is dependent on organizational structure and hierarchy.

Question 2:

As can be seen from Fig. 11.3 of Chapter 11, information is usually generated at the lower levels of an organization and used at the upper levels for decision making. Thus information flows upward through the hierarchy of an organization. Decisions, on the other hand, are sent downward for implementation. This apparent anomaly can have a detrimental impact on the morale and effectiveness of an organization. In construction, for example, it may be time-consuming to wait for a decision to come from the head office, whereas managers at the site office or field office could have been empowered to make necessary decisions based on the available information. Thus the solution is - "Decentralize the decision making authority while centralizing the source of information." The idea is, if consistent, uniform and reliable information is shared, decisions based on that information will be consistent and uniform as well, no matter who is making the decisions and where they are being made.

Question 3:

Data used and generated in the course of a construction project can be grouped under four major categories as follows:

(a) Financial data: cost and pricing, payroll, purchase records, bank statements, income statements and balance sheets.
(b) Administrative data: employee records, contractor records, subcontractors records, suppliers/vendors list, etc.

(c) Field data: status report, materials procurement and inventory, equipment utilization, and time cards.

(d) Technical data: codes, specification, drawings, designs, contracts, etc.

Some of this data are gathered from published sources, such as codes and specifications. Others are generated, such as the design data for a specific project.

Question 4:

Data are communicated using one, or a combination, of the following ways:

- Written formal — e.g., contract documents

- Written informal — e.g., request for information

- Oral formal — e.g., meeting proceedings, verbal directives in meetings

- Oral informal — e.g., onsite verbal communication on project status

One should keep in mind that effectiveness of project management depends on the effectiveness of communication. Formal communication is not always essential - sometimes formality gives rise to unnecessary bureaucracy causing inefficiency. But at other times it is very important to keep proper records of communication, particularly if there are legal implications and consequences. A good project management system would establish rules and procedures outlining the ways of communicating data and information depending on the situation.

ICT tools can be used to enhance all four ways of data communication. For instance, electronic document files containing, texts, images, graphics, sounds, and video footages can be used for all sorts of, formal or informal, written communications. Internet technology and emails as well as fax machines can be very effectively used to transmit this data. Scanning devices are making this technology even more effective. Oral communications, both formal and informal, can be enhanced by the use of, video-conferencing, webcam technology, cellular phones, and various recording devices.

Chapter 12

Question 1:

Planning of a construction project involves several decision-making scenarios. First of all, scope of the project must be defined and it may involve decisions regarding size, location and type of project. Investment decisions are also parts of the planning process. After these major decisions are made, cost engineers, estimators and scheduling engineers must decide on the type of contracts to be used, the techniques of construction to be employed, the sequence of activities to be scheduled, and so forth. Thus planning activities typically involve decision-making at different levels.

Question 2:

During estimating of a construction project, estimators must decide on the techniques of construction to be employed and the size and mix of the crews to be used. The choices on the techniques and the crews must be weighed against time (schedule), cost (price) and quality (specifications). See Example 12.1 of Chapter 12 for a typical decision making scenario during estimating.

Question 3:

There are four major categories of tools used for scheduling. They are described below:

(a) Bar charts

A bar chart graphically represents a project's activities or jobs and their timing. The activities in a bar chart are generally listed on the left side of the chart, vertically one below the other (along the y-axis). A horizontal time scale extends to the right of this list (along the x-axis). A horizontal bar is drawn against each activity listed in the chart between the corresponding start and finish times of the activity.

One of the major advantages of bar charts over other planning and scheduling tools is their relative simplicity. It is quite easy to develop and understand a bar chart. They are, however, very cumbersome to use when the number of activities to be represented becomes large. If several sheets are required, logical interconnections between activities are difficult to comprehend.

(b) Progress curves

Progress curves graphically represent some measure of cumulative progress on the vertical axis against time on the horizontal axis. The progress can be measured in terms of money spent, work-in-place, man-hours expended etc. The units for these cumulative measures of progress can be absolute units (dollars, cubic meters, etc.) or a percentage of the estimated total quantity. In a typical project, resource spending starts slowly, builds to a peak and then tapers off near the end of the project. This results in the cumulative curve representing expenditure of a resource to have a relatively small slope at the start, increase during the middle phases of the project, then flatten out towards the end of the project giving the familiar S - shape to the curve.

(c) Linear balance charts

Linear balance charts or, as they are sometimes called, line of balance (LOB) control charts are best applied to linear and repetitive operations such as tunnels, pipelines, highways and high-rise structures. The vertical axis typically plots cumulative progress, in terms of units completed/produced or percentage completed for different activities or systems of a project. The horizontal axis plots time. The sloping lines represent various activities. The project can be considered to be in good shape as long as the slopes of these lines remain parallel to each other as they move to the right. However, if one activity is proceeding too rapidly with a steep slope compared with that of the activity preceding it then time and space conflicts may result. In other words, if one activity is slower, with a flatter slope, compared with its succeeding activities, there will be potential for conflicts.

(d) Networks

CPM or the critical path method is the most common network technique used to represent activities in a network. Relationships among the activities in terms of dependency can be represented in the network. The technique identifies a project critical path activities on which cannot be delayed without increasing the overall project duration.

Networks can be drawn using one of the two techniques: (a) activity on arrow (AOA), and (b) activity on node (AON). The second technique is also known as the precedence method. Most commercially available project management software tools are based on the precedence method since more complex calculations can be performed using this method than the AOA technique.

There is another variation of the basic CPM technique. It is known as

ABOUT

THE

AUTHORS

S.L. Tang

S.L. Tang is a faculty member in the Department of Civil and Structural Engineering of the Hong Kong Polytechnic University. He is a Chartered Civil Engineer and obtained his B.Sc. in Civil Engineering from the University of Hong Kong in 1972, M.Sc. in Construction Engineering from the National University of Singapore in 1977, and Ph.D. from the Civil Engineering Department of Loughborough University, U.K. in 1989. Dr Tang had about seven years of working experience in civil engineering practice in contracting/consulting firms and government departments before he joined the Hong Kong Polytechnic University. He is currently involved in the teaching and research of construction management, and water and environmental management, and has written over ninety journal/conference papers and books related to the area of his expertise. As a visiting scholar, Dr Tang taught in the civil engineering and construction management departments of several universities, such as University of Technology Sidney of Australia, Tsinghua University, Tongji University, Zhejiang University, and Huaqiao University of China, and Florida International University of USA.

Irtishad U. Ahmad

Irtishad U. Ahmad, P.E. is the Chair of the Department of Construction Management at the Florida International University (FIU) since January 2004. He was the Graduate Program Director in the Department of Civil and Environmental Engineering at FIU from 1998 to 2003. He is a registered professional engineer in the state of Florida. Dr. Ahmad earned his Ph.D. in Civil Engineering from the University of Cincinnati, Ohio, USA in 1988. He taught previously in North Dakota State University, University of Cincinnati, and Bangladesh University of Engineering and Technology. Dr Ahmad is a consultant and a practising engineer in the areas of structural analysis and design, reinforced concrete design, and forensic engineering. He has served in several value engineering teams as a cost engineering and project management expert. He is a frequent speaker in national and international conferences on topics ranging from information technology in construction to effects of wind loading on structures. Dr Ahmad completed several funded research projects related to construction and he published extensively. He is the current Editor-in-Chief of the Journal of Management in Engineering published by the American Society of Civil Engineers.

Syed M. Ahmed

Syed M. Ahmed is a faculty member and Graduate Program Director in the Department of Construction Management at the Florida International University in Miami, Florida USA since December, 1999. Prior to this he was an academic staff member in the Department of Civil & Structural Engineering of the Hong Kong Polytechnic University for over four years. He has a B.Sc. (Hons.) degree in Civil Engineering from the University of Engineering and Technology, Lahore in 1984, and M.S. and Ph.D. degrees in Civil Engineering (majoring in Construction Management with a minor in Industrial and Systems Engineering) from the Georgia Institute of Technology, Atlanta, Georgia, USA in 1989 and 1993 respectively. Dr Ahmed's research interest is in the areas of total quality management, construction safety and quality, contract management and risk analysis, construction procurement and financial management, information technology, and construction and engineering education. He has published extensively in refereed international journals and conferences and has over ten years of construction industry experience.

Ming Lu

Ming Lu is a faculty member in the Department of Civil and Structural Engineering of the Hong Kong Polytechnic University. He has working experience on the planning and construction of numerous civil and industrial projects in the past. In 2000, he earned his Ph.D. degree in Construction Engineering & Management from the University of Alberta, Canada. His current research interest is in the area of computational construction project management aided by artificial intelligence and computer simulations. In recent years, Dr. Lu has developed the novel approaches for neural networks modelling, resource scheduling and operations simulation for enhancing efficiency and productivity of complex construction systems. These developments have been published in the ASCE journals on construction engineering, management, and computing methods.